Six-Minute Solutions

for Civil PE Exam

Geotechnical Depth Problems

Third Edition

Bruce A. Wolle, MSE, PE

Professional Publications, Inc. • Belmont, California

Benefit by Registering This Book with PPI

- Get book updates and corrections.
- Hear the latest exam news.
- Obtain exclusive exam tips and strategies.
- Receive special discounts.

Register your book at **ppi2pass.com/register**.

Report Errors and View Corrections for This Book

PPI is grateful to every reader who notifies us of a possible error. Your feedback allows us to improve the quality and accuracy of our products. You can report errata and view corrections at **ppi2pass.com/errata**.

SIX-MINUTE SOLUTIONS FOR CIVIL PE EXAM GEOTECHNICAL DEPTH PROBLEMS
Third Edition

Current release of this edition: 2

Release History

date	edition number	revision number	update
Jun 2013	2	5	Minor corrections.
Mar 2015	3	1	New edition. Title change. New content and reorganization. Copyright update.
Jul 2016	3	2	Minor corrections. Minor formatting and pagination changes.

Printed in the United States of America.

PPI
1250 Fifth Avenue, Belmont, CA 94002
(650) 593-9119
ppi2pass.com

Library of Congress Control Number: 2015931136

F E D C B A

Table of Contents

TOPIC IX: Shallow Foundations

TOPIC X: Deep Foundations

About the Author

Bruce A. Wolle, MSE, PE, a licensed engineer in 22 states, provides professional engineering services in geotechnical and materials engineering. For more than 25 years he has provided senior engineering supervision and management for projects in the fields of surveying, field explorations and drilling, seismic damage characterizations, design recommendations, and construction quality control/quality assurance in renewable energy development. Mr. Wolle has over 15 years of experience as a civil engineering review course supervisor, coordinator, and instructor. He holds a bachelor of science degree in civil engineering from New Mexico State University and a master of science in engineering degree in civil engineering (geotechnical and seismic studies) from Arizona State University.

Preface and Acknowledgments

Most of my students who passed their exams have told me that solving problems is an invaluable way to become familiar with the exam scope and format and to gain confidence retrieving information from both memory and reference materials. And while professional practice and collegiate texts are sources of varied engineering problems, they are no substitute for a collection of exam-like problems. The problems in this book have been carefully prepared and reviewed to ensure that they cover the civil PE exam geotechnical depth specifications designated by the National Council of Examiners for Engineering and Surveying (NCEES). Although this book doesn't contain any problems from an actual NCEES exam, its problems closely reflect the content, format, and level of difficulty of the problems encountered on the actual exam. The solutions have been reviewed and edited for both accuracy and thoroughness, so you can check your work and learn from any differences between your solving approach and what is presented in this book.

I wrote *Six-Minute Solutions for Civil PE Exam Geotechnical Depth Problems* to provide problem solving practice consistent with the NCEES exam specifications for the geotechnical depth section of the civil PE exam. For the third edition of this book, I included new problems to conform to *ASCE/SEI7: Minimum Design Loads for Buildings and Other Structures* and OSHA 29 CFR 1926. The third edition has been reorganized to align with the topics and distribution of problems on the exam.

Becoming professionally registered as a civil engineer requires dedication and hard work, two fundamental traits of a successful civil engineer. They are also the traits required to produce a book. This book could not have been possible without the dedication and hard work of a number of people. I'd like to thank my countless past students who, through their comments and questions, showed me how solving exam-like problems is necessary for successful exam review. I appreciate the support and guidance of my many colleagues in my efforts to produce exam review materials that safeguard the standard of care and competent practice of our profession. Like this book, engineering is a collaborative effort.

From PPI, I acknowledge the significant editorial contributions of Sarah Hubbard, director of product development and implementation. Thank you also to the rest of the product development and implementation staff: Cathy Schrott, production services manager; Nicole Evans, EIT, associate project manager and staff engineer; Heather Turbeville, associate project manager; Ian A. Walker, lead editor; Tom Bergstrom, production associate and technical illustrator; Kate Hayes, production associate; David Chu, Tyler Hayes, and Julia Lopez, copy editors; Ralph Arcena, EIT, engineering intern; and Jennifer Lindeburg King, associate editor-in-chief.

I would also like to thank technical reviewer Charles Hubbard, who technically reviewed the new problems in this edition.

I am grateful for the support of my wife, Tammy, and my family, friends, and colleagues. Without their support and patience, this book would not have been possible.

Problems and solutions have been carefully prepared and reviewed to ensure that they are appropriate and understandable, and that they were solved correctly. If you find errors or discover an alternative, more efficient way to solve a problem, please bring it to PPI's attention so your suggestions can be incorporated into future editions. You can report errors and keep up with the changes made to this book by logging on to PPI's website at **ppi2pass.com/errata**.

Thanks, and best of luck in your future endeavors!

Bruce A. Wolle, MSE, PE

Introduction

ABOUT THIS BOOK

Six-Minute Solutions for Civil PE Exam Geotechnical Depth Problems is organized into 10 chapters. Each chapter contains problems that correspond to the format and scope that would be expected to appear in the civil PE exam's geotechnical depth section.

Most of the problems are quantitative, requiring calculations to arrive at a correct solution. A few are nonquantitative. Some problems will require a little more than 6 minutes to answer, and others require a little less. On average, during the exam you should expect to complete 80 problems in 480 minutes (8 hours), or spend 6 minutes per problem.

All of the problems include a hint to provide direction in solving the problem. The solutions are presented in a step-by-step sequence to help you follow the logical development of the correct solution and to provide examples of how you may want to approach your solutions as you take the PE exam.

In addition to the correct solution, you will find an explanation of the faulty solutions leading to the three incorrect answer options. The incorrect options are intended to represent common mistakes specific to different problem types. These may be simple mathematical errors, such as failing to square a term in an equation, or more serious errors, such as using the wrong equation.

Though you will not encounter problems on the exam exactly like those presented in this book, reviewing these problems and solutions will increase your familiarity with the exam problems' format, content, and solution methods. This preparation will help you considerably during the exam.

ABOUT THE EXAM

The civil PE exam is divided into two four-hour sessions, with each session containing 40 multiple-choice problems each. Only one of the four options given is correct, and the problems are completely independent of each other.

The morning section of the civil PE exam is a "breadth" exam covering eight areas of general civil engineering knowledge: project planning, means and methods, soil mechanics, structural mechanics, hydraulics and hydrology, geometrics, materials, and site development. All examinees take the same morning exam.

You must choose one of the five afternoon exam sections: construction, geotechnical, structural, transportation, or water resources and environmental. The geotechnical depth section is intended to assess your knowledge of geotechnical engineering principles and practices. The topics and approximate distribution of problems for the geotechnical depth section are as follows.

1. **Site Characterization (5 questions):** Interpretation of available existing site data and proposed site development data; subsurface exploration planning; geophysics; drilling techniques; sampling techniques; in situ testing; description and classification of soils; rock classification and characterization; groundwater exploration, sampling, and characterization

2. **Soil Mechanics, Laboratory Testing, and Analysis (5 questions):** Index properties and testing; strength testing of soil and rock; stress-strain testing of soil and rock; permeability testing properties of soil and rock; effective and total stresses

3. **Field Materials Testing, Methods, and Safety (3 questions):** Excavation and embankment, borrow source studies, laboratory and field compaction; trench and construction safety; geotechnical instrumentation

4. **Earthquake Engineering and Dynamic Loads (2 questions):** Liquefaction analysis and mitigation techniques; seismic site characterization, including site classification using ASCE 7; pseudo-static analysis and earthquake loads

5. **Earth Structures (4 questions):** Slab on grade; ground improvement; geosynthetic applications; slope stability and slope stabilization; earth dams, levees, and embankments; landfills and caps; pavement structures (rigid, flexible, or unpaved), including equivalent single-axle load (ESAL), pavement thickness, subgrade testing, subgrade preparation, maintenance and rehabilitation treatments; settlement

6. **Groundwater and Seepage (3 questions):** Seepage analysis/groundwater flow; dewatering design, methods, and impact on nearby

structures; drainage design/infiltration; grouting and other methods of reducing seepage

7. **Problematic Soil and Rock Conditions (3 questions):** Karst; collapsible, expansive, and sensitive soils; reactive/corrosive soils; frost susceptibility

8. **Earth Retaining Structures (ASD or LRFD) (5 questions):** Lateral earth pressure; load distribution; rigid retaining wall stability analysis; flexible retaining wall stability analysis; cofferdams; underpinning; ground anchors, tie-backs, soil nails, and rock anchors for foundations and slopes

9. **Shallow Foundations (ASD or LRFD) (5 questions):** Bearing capacity; settlement, including vertical stress distribution

10. **Deep Foundations (ASD or LRFD) (5 questions):** Single-element axial capacity; lateral load and deformation analysis; single-element settlement; downdrag; group effects; installation methods/hammer selection; pile dynamics; pile and drilled-shaft load testing; integrity testing methods

HOW TO USE THIS BOOK

To optimize your study time and obtain the maximum benefit from these problems, consider the following suggestions.

1. Complete an overall review of the problems, and identify the subjects that you are least familiar with. Work a few of these problems to assess your general understanding of the subjects and to identify your strengths and weaknesses.

2. Locate and organize relevant resource materials. (See the References section in this book as a starting point.) As you work problems, some of these resources will emerge as more useful to you than others. These are what you will want to have on hand when taking the PE exam.

3. Work the problems in one chapter at a time, starting with the subject areas that you have the most difficulty with.

4. When possible, work problems without utilizing the hints. Always attempt your own solutions before looking at the solutions provided in the book. Use the solutions to check your work or to provide guidance in solving the more difficult problems. Use the incorrect solutions to help identify pitfalls and to develop strategies to avoid them.

5. Use each chapter's solutions as a guide to understanding general problem-solving approaches. Although problems identical to those presented in *Six-Minute Solutions for Civil PE Exam Geotechnical Depth Problems* will not be encountered on the PE exam, the approach to solving problems will be similar.

For further information and tips on how to prepare for the civil PE exam's geotechnical depth section, consult the *Civil Engineering Reference Manual* or PPI's website, **ppi2pass.com/cefaq**.

Codes and References Used to Prepare This Book

The following codes and references were used to prepare this book. The listed codes have also been adopted as design standards for the geotechnical depth section of the civil PE exam. The information used to write and update this book was based on exam specifications at the time of publication. However, as with engineering practice itself, the PE exam is not always based on the most current codes or cutting-edge technology. Similarly, codes, standards, and regulations adopted by state and local agencies often lag issuance by several years. It is likely that the codes that are most current, the codes that you use in practice, and the codes that are the basis of your exam will all be different.

PPI lists on its website the dates and editions of the codes, standards, and regulations on which NCEES has based the PE exams (**ppi2pass.com/cefaq**). It is your responsibility to find out which codes will be tested on your exam.

CODES

ASCE/SEI7: *Minimum Design Loads for Buildings and Other Structures*, 2010, American Society of Civil Engineers, Reston, VA

OSHA: *Safety and Health Regulations for Construction*, 29 CFR Part 1926 (U.S. Federal version), U.S. Department of Labor, Washington, DC

REFERENCES

Bowles, Joseph E. *Foundation Analysis and Design*. New York, NY: McGraw-Hill.

Das, Braja M. and Khaled Sobhan. *Principles of Geotechnical Engineering*. Stamford, CT: Cengage Learning.

Holtz, Robert D., William D. Kovacs, and Thomas C. Sheahan. *An Introduction to Geotechnical Engineering*. Upper Saddle River, NJ: Prentice Hall.

Kramer, Steven L. *Geotechnical Earthquake Engineering*. Upper Saddle River, NJ: Prentice Hall.

Reese, Lymon C., and William F. Van Impe. *Single Piles and Pile Groups Under Lateral Loading*. Boca Raton, FL: CRC Press.

Youd, T. L., I. M. Idriss, R. D. Andrus, et. al. "Liquefaction Resistance of Soils: Summary Report from the 1996 NCEER and 1998 NCEER/NSF Workshops on Evaluation of Liquefaction Resistance of Soils." *Journal of Geotechnical and Geoenvironmental Engineering*, Vol. 127, No. 10. American Society of Civil Engineers.

Nomenclature

a	acceleration	ft/sec^2	m/s^2
a	constant coefficient for quadratic equation	–	–
A	amplitude	in	mm
A	annual amount	–	–
A	area	ft^2	m^2
AADT	average annual daily traffic	veh/day	veh/d
ADT	average daily traffic	veh/day	veh/d
b	constant coefficient for quadratic equation	–	–
b	group width	ft	m
b	strip width	in	m
B	boring	–	–
B	footing or foundation element width	ft	m
c	cohesion	lbf/ft^2	Pa
c	constant coefficient for quadratic equation	–	–
c	undrained shear strength (cohesion)	lbf/ft^2	Pa
c_A	adhesion	lbf/ft^2	Pa
C	circumference	ft	m
C	correction factor	–	–
C_c	compression index	–	–
C_Q	factor for normalizing CPT tip penetration	–	–
C_r	compression ratio	–	–
C_r	recompression index	–	–
C_u	uniformity coefficient	–	–
C_v	coefficient of consolidation	–	–
C_z	coefficient of curvature	–	–
CBR	California bearing ratio	–	–
CPT	tip penetration	–	–
CSR	cyclic stress ratio	–	–
d	depth factor, D/H	–	–
d	diameter	ft	m
d	distance	ft	m
D	depth	ft	m
D	diameter	ft	m
D	drainage coefficient	–	–
e	void ratio	–	–
E	hammer energy	ft-kips	N·m
E	modulus of elasticity	lbf/in^2	Pa
E_s	Young's modulus	lbf/in^2	Pa
ESAL	equivalent single-axle loads	–	–
f	coefficient of friction	–	–
f_R	friction ratio	%	%
F	factor of safety	–	–
F	normalized friction ratio	–	–
g	gravitational acceleration	ft/sec^2	m/s^2
g	gravitational constant	ft-lbm/lbf-sec^2	n.a.
g	growth rate per period	decimal, % per unit time	decimal, % per unit time
G	universal gravitational constant	lbf-ft^2/lbm^2	N·m^2/kg^2
GF	growth-rate factor	–	–
h	depth below the water table	ft	m
h	height or thickness	ft	m
h	horizontal	–	–
H	elevation difference	ft	m
H	height or thickness	ft	m
i	effective rate per period (usually per year)	decimal per unit time	decimal per unit time
i	hydraulic gradient	–	–
I	moment of inertia	ft^4	m^4
I	interest per period	% per unit time	% per unit time
I_c	soil behavior type index	–	–
I_g	group index	–	–
k	earth pressure coefficient	–	–
k	modulus of subgrade reaction	lbf/in^3	N/m^3
K	coefficient of permeability	ft/sec	m/s
K	earth pressure constant	–	–
K	relative stiffness	–	–
K_s	correction for grain characteristics	–	–
l	length	ft	m
L	length	ft	m
LL	liquid limit	%	%
m	drainage coefficient	–	–
m	mass	lbm	g
M	earthquake Richter magnitude	–	–
M	moment	ft-lbf	N·m
M_R	resilient modulus	lbf/in^2	Pa

M_R	resisting moment of wall	in-lbf	N·m
n	iteration exponent for liquefaction analysis	–	–
n	number	%	%
n	porosity	–	–
N	bearing capacity factor	–	–
N	number	–	–
N	standard penetration resistance	blows/ft	blows/m
$\overline{N}$	average standard penetration resistance	blows/ft	blows/m
N_o	stability number	–	–
OCR	overconsolidation ratio	–	–
OM	optimum moisture content	%	%
p	lateral force per unit length from pile	lbf/in	N/m
p	perimeter	ft	m
p	pressure	lbf/ft^2	Pa
p_i	effective pressure	lbf/ft^2	Pa
P	load	kips	N
P	permeability	various	various
PI	plasticity index	–	–
PL	plastic limit	–	–
q	bearing capacity	lbf/ft^2	Pa
q	compressive strength	lbf/ft^2	Pa
q	pressure under footing	lbf/ft^2	Pa
q	surcharge	lbf/ft^2	Pa
q_c	tip resistance	lbf/ft^2	Pa
q_s	sleeve friction resistance	lbf/ft^2	Pa
Q	capacity	lbf	N
Q	CPT soil behavior type index	–	–
Q	flow quantity or rate	ft^3/sec	m^3/s
Q_s	strength reduction factor	–	–
r	radius	ft	m
r	ratio	–	–
R	earth pressure resultant	lbf	N
R	radius	ft	m
R	ratio	–	–
R	rigidity	–	–
RR	recompression ratio	–	–
s	spacing	in	mm
s	stress	lbf/in^2	Pa
S	degree of saturation	%	%
S	distance per hammer blow	in	m
S	settlement	ft	m
S	shape factor	–	–
S	shear strength	lbf/ft^2	Pa
S	spectral acceleration	ft/sec^2	m/s^2
S	strength	lbf/in^2	Pa
SG	specific gravity	–	–
SN	structural number	–	–
t	time	sec	s
t	thickness	ft	m
T	tensile force	lbf	N
T_v	time factor	–	–
u	pore pressure	lbf/ft^2	Pa
U_z	degree of consolidation	–	–
v	velocity	in/sec	m/s
V	base shear	lbf	N
V	volume	ft^3	m^3
w	average axle loading	lbf	N
w	group length	ft	m
w	moisture content	%	%
w	width	ft	m
w_{18}	equivalent single-axle load for design lane	–	–
$\widehat{w}_{18}$	total equivalent single-axle load for design lane	–	–
W	vertical force (weight)	lbf	N
x	fraction by weight	–	–
x	quadratic root	–	–
y	vertical displacement	in	mm

Symbols

α	adhesion factor	–	–
α	angle	deg, rad	deg, rad
α	coefficient (liquefaction)	–	–
α	coefficient of compression	–	–
β	angle or slope	deg	deg
β	coefficient (liquefaction)	–	–
β	effective stress factor	–	–
γ	specific weight or unit weight	lbf/ft^3	N/m^3
δ	friction angle	–	–
Δ	deflection	ft	m
Δ	deformation	ft	m
Δ	displacement	ft	m
Δ	elongation	in	mm
ε	eccentricity	ft	m
ε	strain	–	–
θ	angle	deg, rad	deg, rad
λ	rake angle of retaining wall face	deg	deg
μ	pore pressure	lbf/ft^2	Pa
ν	Poisson's ratio	–	–
ρ	density	lbm/ft^3	kg/m^3
σ	normal stress	lbf/ft^2	Pa
τ	shear stress	lbf/ft^2	Pa
υ	specific volume	ft^3/lbm	m^3/kg
ϕ	angle of internal friction	deg	deg
ϕ'	effective angle of internal friction	deg	deg

Subscripts

γ	density or unit weight
0	initial
a	active, air, allowable, area, or axial
add	additional
atm	atmospheric
ave	average
A	adhesion or axial
b	buoyant
c	cell, cohesive, cone tip, container, or correction
d	drainage, dry, or equipotential drop
dsc	dry soil plus container
eff	effective
emb	embankment
f	failure, field, final, flow channel, or footing
fs	field sample
g	air (gas)
h	horizontal
i	initial, inner, or inside
l	length
L	length
m	moist
max	maximum
min	minimum
n	blow count, nominal, normal, or period n
o	at rest, center-to-center, initial, original, out, outer, outlet, outside, or overall
O	outside
OT	overturning
p	bearing, equipotential, passive, pile tip, potential, or pressure
q	surcharge
Q	from unit load
r	radius, ratio, recompression, or resultant
R	Rankine, resistance, resisting, resistive, or resultant
s	side, skin, soil, solid, specific, or static
sat	saturated
set	settlement
SL	sliding
sr	single ring
ss	suction specific
sub	submerged
t	thickness, time, or total
u	unconfined or undrained
uc	unconfined compression
ult	ultimate
ut	ultimate tensile
v	vertical, void, or volumetric
w	wall, water, or width
wsc	wet soil plus container
WT	water table
x	in x-direction
y	in y-direction or yield
z	at depth z or zero air voids

1 Site Characterization

IN SITU TESTING

PROBLEM 1

Commencing from 2 ft below the ground surface and using a standard penetration test (SPT) split-barrel sampler driven 18 in, the following blow-count data was obtained in increments of 6 in: 10, 15, 19. What is the N-value to be reported on the boring log for the depth of 3 ft?

(A) 19 blows/ft

(B) 25 blows/ft

(C) 34 blows/ft

(D) 44 blows/ft

Hint: In a standard penetration test (SPT), a seating interval of 6 in precedes the actual blow-count measurement.

PROBLEM 2

Field standard penetration test data is obtained for depths ranging from 15 ft to 16.5 ft below the ground surface where the groundwater level is at 8 ft below the ground surface. Boring logs and laboratory testing indicate the soil profile consists of a silty sand layer approximately 20 ft thick with an average dry unit weight of 117.7 lbf/ft^3. The average water content was determined to be 3.8% above the water table and 14.2% below the water table.

depth (ft)	N-count (blows/ft)
15–15.5	12
15.5–16	15
16–16.5	16

Determine the corrected N-value.

(A) 31 blows/ft

(B) 35 blows/ft

(C) 37 blows/ft

(D) 50 blows/ft

Hint: The N-value should be corrected to a value expected for a standard effective overburden stress of 2000 lbf/ft^2.

INTERPRETATION OF SITE DATA

PROBLEM 3

The soil test boring shown was presented in a geotechnical report for a proposed hotel. The test borings were advanced within the proposed foundation perimeter of the hotel. The total load per interior column (net dead plus live) is expected to be approximately 100 kips. A maximum differential settlement of 1 in is acceptable.

depth (ft)	soil classification and remarks	soil symbol	elevation (ft)	sampler type	N-count (1st 6 in, 2nd 6 in, 3rd 6 in)
0	brown and black silty CLAY (USCS: CH), very soft, medium to high moisture content, high plasticity		150	SS SS	4–6–6 5–6–7
10	dark brown sandy CLAY (USCS: CL), very soft to soft, high moisture content, medium to high plasticity		140	SS SS SS	4–3–4 4–5–6 6–8–9
20	dark brown/gray clayey SAND (USCS: SC), dense to very dense, high moisture content, low to medium plasticity		130	SS SS	9–9–11 12–14–19
30	brown/gray silty GRAVEL (USCS: GM) with clay and sand, very dense to hard, high moisture content, low to medium plasticity		120	SS SS	25–32–33 25–50/3 in

Assuming a local contractor is available to construct each type, and including economical considerations, what is the best recommendation for the foundation?

(A) deep spread footings

(B) drilled piers or driven piles

(C) mat or raft

(D) slab-on-grade

Hint: Consider the presence of shallow groundwater with soft clay and low blow counts, along with the expected loading conditions and acceptable differential settlement.

PROBLEM 4

A field exploration was performed at a site consisting of rural vacant land for a proposed monopole telecommunications tower. The required base reactions for foundation design given by the tower manufacturer are shown.

tower weight	13.25 kips
total shear	8.74 kips
overturning moment	433 ft-kips

The test boring record shown was developed based on the field exploration.

depth (ft)	soil classification and remarks	soil symbol	elevation (ft)	sampler type	*N*-count 1st 6 in 2nd 6 in 3rd 6 in
0	tan/brown silty SAND (USCS: SM), medium dense, low moisture, low plasticity, white to gray with tan brown streaks, strongly cemented, very dense		420	SS SS SS	12–14–18 24–32–34 18–23–29
10	brown clayey SAND (USCS: SC) with gravel, medium dense to dense, medium moisture, low to medium plasticity		410	SS SS	15–17–18 17–19–21
20	brown to dark brown poorly graded GRAVEL (USCS: GP), dense to very dense, medium to high moisture, medium plasticity		400	SS SS	24–24–24 29–32–37
30	brown to gray silty GRAVEL (USCS: GM) with sand and cobbles, very dense, medium to high moisture, low to medium plasticity		390	SS SS	32–42–45 46–50/4 in

What type of foundation is most feasible?

(A) drilled cast-in-place pier/caisson

(B) driven precast concrete piles

(C) mat-and-pedestal

(D) thickened slab-on-grade

Hint: Consider the subsurface soil conditions and the feasibility to construct each option.

PROBLEM 5

A mat-and-pedestal foundation will be designed for a tower.

depth (ft)	soil classification and remarks	soil symbol	elevation (ft)	sampler type	*N*-count 1st 6 in 2nd 6 in 3rd 6 in
0	tan/brown silty SAND (USCS: SM), medium dense, low moisture, low plasticity, white to gray with tan brown streaks, strongly cemented, very dense		420	SS SS SS	12–14–18 24–32–34 18–23–29
10	brown clayey SAND (USCS: SC) with gravel, medium dense to dense, medium moisture, low to medium plasticity		410	SS SS	15–17–18 17–19–21
20	brown to dark brown poorly graded GRAVEL (USCS: GP), dense to very dense, medium to high moisture, medium plasticity		400	SS SS	24–24–24 29–32–37
30	brown to gray silty GRAVEL (USCS: GM) with sand and cobbles, very dense, medium to high moisture, low to medium plasticity		390	SS SS	32–42–45 46–50/4 in

Using the test boring information shown, what would be the governing engineering condition for placement depth of the concrete mat?

(A) bearing capacity

(B) lateral resistance

(C) overturning resistance

(D) allowable settlement

Hint: Based on the observed subsurface soil conditions, consider what base reaction criteria will have the most influence on establishing the appropriate factor of safety for the entire structure.

SAMPLING TECHNIQUES

PROBLEM 6

A drilling subcontractor proposes to use a standard split-spoon penetration sampler for a geotechnical field exploration project. The dimensions of the sampler are

1.75 in outside diameter and 1.63 in cutting-edge diameter. The area ratio of the sampler is most nearly

(A) 7.3%

(B) 15%

(C) 25%

(D) 88%

Hint: The area ratio is expressed as the ratio of the volume of soil displacement to the volume of the collected sample.

SOLUTION 1

The N-value is obtained by counting the number of blows needed to drive the sampler for 12 in after a 6 in seating drive. The blow counts for the final two sets of 6 in increments are summed to get the resulting N-value.

$$\begin{aligned} N_{\text{value}} &= N_{\text{2nd 6 in}} + N_{\text{3rd 6 in}} \\ &= 15\ \frac{\text{blows}}{\text{ft}} + 19\ \frac{\text{blows}}{\text{ft}} \\ &= 34\ \text{blows/ft} \end{aligned}$$

The answer is (C).

Why Other Options Are Wrong

(A) This incorrect solution is the blow count recorded during the final 6 in increment.

(B) This incorrect solution is the sum of the blow counts from the first two sets of 6 in increments instead of from the final two sets.

(D) This incorrect solution is the sum of the blow count values for the entire 18 in drive.

SOLUTION 2

The field N-value can be determined from the field data.

$$\begin{aligned} N_{\text{value,field}} &= N_{\text{2nd 6 in}} + N_{\text{3rd 6 in}} \\ &= 15\ \frac{\text{blows}}{\text{ft}} + 16\ \frac{\text{blows}}{\text{ft}} \\ &= 31\ \text{blows/ft} \end{aligned}$$

The standardized N-value can be determined from the field N-value using

$$N_{\text{value,corrected}} = C_N N_{\text{value,field}}$$

The typical procedure is to adjust the field N-value in reference to a standard overburden stress of 1 ton/ft^2 (2000 lbf/ft^2).

$$C_N = \sqrt{\frac{\sigma_{v,\text{ref}}}{\sigma_v'}}$$

The effective stress at the blow count depth of 16 ft is used in the calculation.

$$\begin{aligned}\sigma'_v &= \gamma_d(1+w_{\text{above WT}})H_{\text{above WT}} \\ &\quad +(\gamma_d(1+w_{\text{below WT}})-\gamma_w)H_{\text{below WT}} \\ &= \left(117.7\ \frac{\text{lbf}}{\text{ft}^3}\right)(1+0.038)(8\ \text{ft}) \\ &\quad +\left(\left(117.7\ \frac{\text{lbf}}{\text{ft}^3}\right)(1+0.142)-62.4\ \frac{\text{lbf}}{\text{ft}^3}\right) \\ &\quad \times(8\ \text{ft}) \\ &= 1553\ \text{lbf/ft}^2\end{aligned}$$

$$\begin{aligned}C_N &= \sqrt{\frac{\sigma_{v,\text{ref}}}{\sigma'_v}} = \sqrt{\frac{1\ \frac{\text{ton}}{\text{ft}^2}}{1553\ \frac{\text{lbf}}{\text{ft}^2}}} = \sqrt{\frac{2000\ \frac{\text{lbf}}{\text{ft}^2}}{1553\ \frac{\text{lbf}}{\text{ft}^2}}} \\ &= 1.13\end{aligned}$$

The corrected N-value can be calculated as

$$\begin{aligned}N_{\text{value,corrected}} &= C_N N_{\text{value,field}} = (1.13)\left(31\ \frac{\text{blows}}{\text{ft}}\right) \\ &= 35\ \text{blows/ft}\end{aligned}$$

The answer is (B).

Why Other Options Are Wrong

(A) This solution erroneously calculates the effective stress by failing to subtract the pore pressure and instead calculating the total stress.

(C) This solution fails to convert the dry unit weight to a total unit weight before calculating the effective stress.

(D) This solution erroneously adds all three N-counts when only the final two increments should be added. The effective stress would then be incorrectly assumed to be calculated at a depth of 15 ft rather than 16 ft.

SOLUTION 3

The test boring record indicates a shallow groundwater table exists approximately 10 ft below the ground surface, with soft clay soils to a depth of approximately 25 ft. The soft clay is underlain by dense sand and by gravel and rock. The use of drilled piers is most desirable, but driven piles may also be feasible from cost considerations. In comparison to the other options, drilled piers can be constructed in the presence of groundwater while providing end-bearing support and resulting in minimal settlement.

The answer is (B).

Why Other Options Are Wrong

(A) This solution is incorrect because this type of foundation requires excavation to a depth sufficient to achieve adequate bearing capacity or would require the use of very large footings. Differential settlement can be expected due to variation in stress distribution from high structural loads. A larger quantity of concrete would be required over the most feasible option, drilled piers or driven piles, thereby also resulting in an expensive option.

(C) This solution is incorrect because a mass excavation would require dewatering, shoring design, and a large quantity of concrete. These additional requirements are more than likely expensive in comparison to drilled piers or driven piles.

(D) This solution is incorrect because this type of structure may or may not provide adequate bearing capacity; it would be subject to excessive differential settlement and cracking. A large quantity of concrete would be required as well, driving up costs over those for drilled piers or driven piles.

SOLUTION 4

The mat-and-pedestal option is the most feasible in consideration of very dense cemented gravel and sand observed approximately 5 ft below the ground surface. Although the soils may tend to ravel, the site is on vacant land, allowing for shallow slopes as needed for a wide excavation.

The answer is (C).

Why Other Options Are Wrong

(A) A drilled pier could be constructed, but this is not the most feasible option. The subsurface would prove extremely difficult to drill due to its composition of very dense, cemented sand and gravel. Furthermore, the sandy soils would ravel and collapse, requiring downhole stabilization techniques such as casing or mud, making this a more expensive and prohibitive option compared to a mat-and-pedestal foundation.

(B) This option is not feasible, since the cemented subsurface conditions have high potential to damage the precast concrete piles during driving.

(D) A thickened slab-on-grade would not provide adequate overturning resistance. An oversized concrete gravity block or similar structure could be designed to resist overturning; however, that option is different than a thickened slab-on-grade and was not given. An oversized thickened slab-on-grade could be used; however, the potential for higher costs due to concrete in comparison to a mat-and-pedestal foundation would be greater.

SOLUTION 5

Although bearing capacity and settlement should be evaluated, the governing condition for design of the mat foundation is overturning resistance. The appropriate surcharge pressure above the mat is required based on placement depth and achievement of compacted density for backfill soils.

The answer is (C).

Why Other Options Are Wrong

(A) This solution is incorrect because the required governing criterion being requested by the problem statement is the depth of installation, not the size of the mat. In consideration of the *N*-values given for the subsurface soils, bearing capacity is not a governing condition for the placement depth. The mat structure could be constructed at the surface or at a relatively shallow depth and should still achieve adequate bearing capacity due to the reported *N*-values.

(B) This solution is incorrect, since lateral resistance should not affect the design of a mat-and-pedestal foundation structure in the types of soils observed. The lateral resistance is provided by the frictional properties of the base of the mat foundation and by soils at the bearing level in the form of sliding resistance. Resistance to sliding would be achieved at the ground surface soils as well as at the soils found at depth.

(D) This solution is incorrect because the subsurface profile consists of dry, very dense sand-and-gravel soils that exhibit immediate and minimal settlement conditions rather than long-term settlement properties characteristic of saturated clay layers.

SOLUTION 6

The area ratio given as a percentage is

$$\begin{aligned} A_r &= \frac{D_o^2 - D_i^2}{D_i^2} \times 100\% \\ &= \frac{(1.75\text{ in})^2 - (1.63\text{ in})^2}{(1.63\text{ in})^2} \times 100\% \\ &= 15\% \end{aligned}$$

The answer is (B).

Why Other Options Are Wrong

(A) This incorrect solution does not square any values. The value is less than the correct answer of 15%.

(C) This incorrect solution squares the terms in the numerator only, but does not square the denominator. The value is greater than the correct answer of 15%.

(D) This incorrect solution squares the nominal outside diameter term only, but does not square the cutting-edge diameter. The value is much greater than the correct answer.

2 Soil Mechanics, Lab Testing, and Analysis

EFFECTIVE AND TOTAL STRESSES

PROBLEM 1

Approximate the effective stress at midpoint in the sandy clay layer given the soil profile shown.

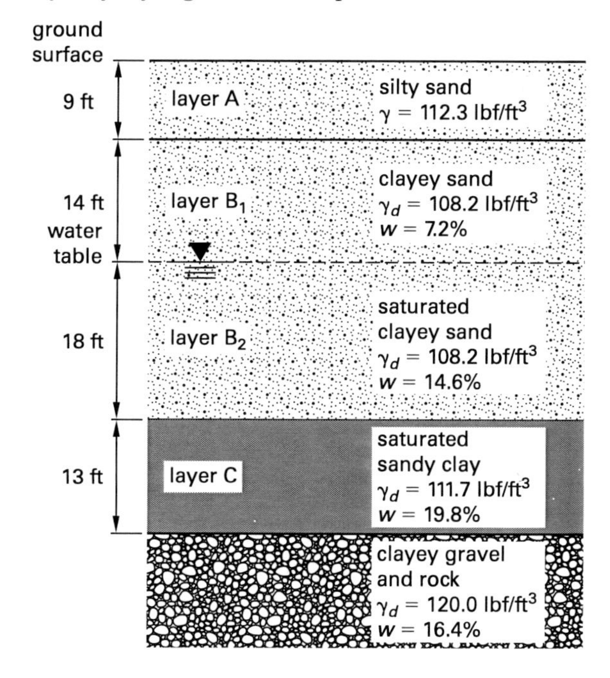

(A) 3670 lbf/ft^2

(B) 4210 lbf/ft^2

(C) 4670 lbf/ft^2

(D) 5740 lbf/ft^2

Hint: In hydrostatic conditions, effective stress equals the pore pressure subtracted from the total stress.

PROBLEM 2

The soil above the water table in the given soil profile is dry.

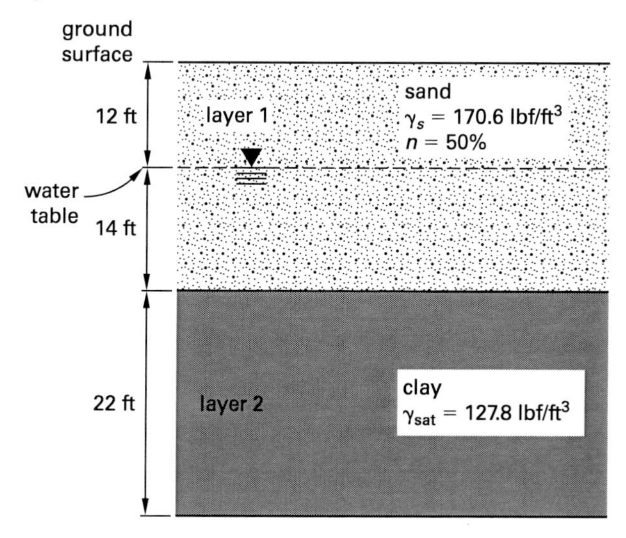

Approximate the effective stress at the base of layer 2.

(A) 3220 lbf/ft^2

(B) 3810 lbf/ft^2

(C) 5440 lbf/ft^2

(D) 5470 lbf/ft^2

Hint: The effective stress equals the pore pressure subtracted from the total stress under hydrostatic conditions. To get started, assume the total volume is a unit value.

PROBLEM 3

Given the soil profile shown, what is most nearly the total horizontal stress at a depth of 23 ft?

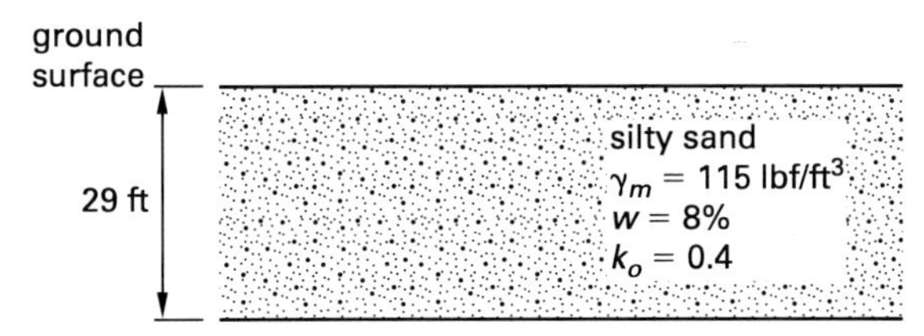

(A) 980 lbf/ft^2

(B) 1060 lbf/ft^2

(C) 1140 lbf/ft^2

(D) 1330 lbf/ft^2

Hint: In the absence of a groundwater table, there is no pore water pressure; hence, the total stress and effective stress are equal.

INDEX PROPERTIES AND TESTING

PROBLEM 4

The following laboratory test results for Atterberg limits and particle-size distribution (sieve analysis) were obtained for a soil.

particle size distribution (percent passing)

sieve no.	B3 (10–11 ft)
no. 4 (4.75 mm)	80
no. 10 (2.00 mm)	60
no. 40 (0.425 mm)	30
no. 200 (0.075 mm)	10
Atterberg limits	
liquid limit	31
plastic limit	25

Classify the soil according to the Unified Soil Classification System (USCS).

(A) SM

(B) SP

(C) SW-SC

(D) SW-SM

Hint: There is no need to graph the given data. Use the coefficient of conformity and the coefficient of curvature to help assess the shape parameter of the gradation curve.

PROBLEM 5

The following laboratory results were obtained for a soil sample.

particle size distribution (percent passing)

sieve no.	B1 (5–6 ft)
no. 8 (2.36 mm)	100.0
no. 16 (1.18 mm)	99.3
no. 30 (0.600 mm)	97.9
no. 50 (0.300 mm)	93.1
no. 100 (0.150 mm)	77.2
no. 200 (0.075 mm)	58.6
Atterberg limits	
liquid limit	29
plastic limit	19

Classify B1 (5–6 ft) according to the AASHTO system.

(A) A-3

(B) A-4

(C) A-6

(D) A-7-6

Hint: Use the AASHTO soil classification flow chart.

PROBLEM 6

Using the data presented for the given soil sample, determine the soil classification using the Unified Soil Classification System (USCS).

particle size distribution (percent passing)	
sieve no.	B3 (10–11 ft)
no. 8 (2.36 mm)	100.0
no. 16 (1.18 mm)	99.4
no. 30 (0.600 mm)	96.8
no. 50 (0.300 mm)	93.0
no. 100 (0.150 mm)	86.0
no. 200 (0.075 mm)	72.0
Atterberg limits	
liquid limit	38
plastic limit	20

(A) CL

(B) CH

(C) ML

(D) MH

Hint: The coefficient of conformity and the coefficient of curvature will not help assess the shape parameter of the gradation curve.

PROBLEM 7

The data shown were obtained during a plastic limit (PL) test.

m_{wsc}	mass of wet soil plus container	23.42 g
m_{dsc}	mass of dry soil plus container	19.81 g
m_{c}	mass of container itself	1.73 g

The PL of the soil is most nearly

(A) 18

(B) 20

(C) 30

(D) 85

Hint: The PL of the soil is the water content of the test sample reported without the percentage symbol.

PROBLEM 8

During compaction of a parking area embankment under construction, a contractor reports the occurrence of pumping and rutting. In a field sand cone test, a 54 lbf soil sample with a volume of 0.44 ft^3 was obtained by filling the hole with dry test sand. Using a field oven, the sample was dried to a new weight of 43 lbf. Given laboratory tests that indicate the specific gravity of the solids is 2.66, what is most nearly the original void ratio of the test sample?

(A) 0.2

(B) 0.3

(C) 0.7

(D) 2

Hint: The void ratio can be determined using a phase relationship because the total volume is known.

PROBLEM 9

A soil sample weighs 54 lbf and has a volume of 0.39 ft^3. The sample is dried to a new weight of 43 lbf. Determine the maximum water content the soil can obtain without swelling or bleeding. The specific gravity of the solids is 2.66.

(A) 15%

(B) 19%

(C) 26%

(D) 38%

Hint: The volume of the voids equals the volume of the water in saturated conditions.

PROBLEM 10

A modified California ring sampler was used in the field to obtain the in-situ moisture content and unit weight of a soil. The recovery was five full rings with a total mass of 879 g. The following data are given.

m_{ring}	average mass of a single ring	44 g
D	average diameter of a single ring	2.42 in
H	average height of a single ring	1 in
w	moisture content of the sample	11.2%

The dry unit weight of the soil is most nearly

(A) 100 lbf/ft^3

(B) 110 lbf/ft^3

(C) 120 lbf/ft^3

(D) 130 lbf/ft^3

Hint: Find the in-situ total density of the soil sample and then convert the result to a dry unit weight.

Soil Mech. Lab Test & Analysis

PROBLEM 11

Given the phase diagram shown for a saturated soil, what is most nearly the void ratio?

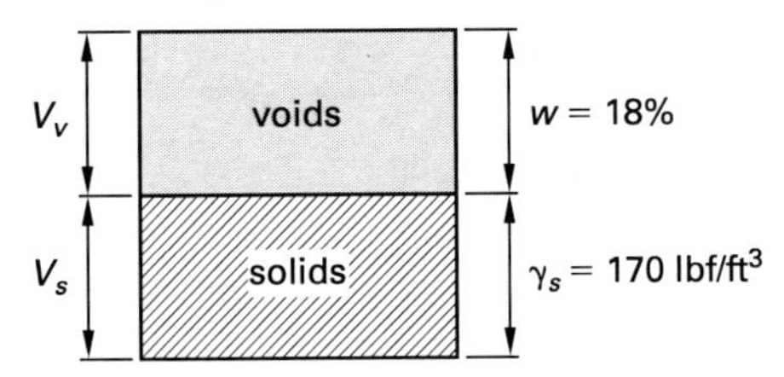

(A) 0.25

(B) 0.49

(C) 1.9

(D) 3.0

Hint: Assume a unit value for the volume of solids.

PROBLEM 12

Given the phase diagram shown, what is most nearly the porosity?

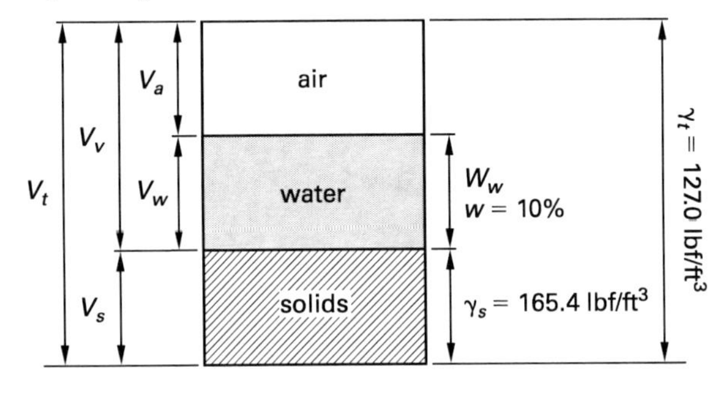

(A) 0.42%

(B) 3.3%

(C) 9.1%

(D) 30%

Hint: Start by determining the weight of water using the water content and weight of solids.

PERMEABILITY

PROBLEM 13

In a constant-head permeability test, a soil sample is compacted into a round PVC pipe and placed into a tank as shown. After sufficient time is allowed for saturation of the soil, the test is conducted under standard temperature-pressure (STP) conditions for 4 h. A total of 227 g of water is decanted into a measuring vessel during the test.

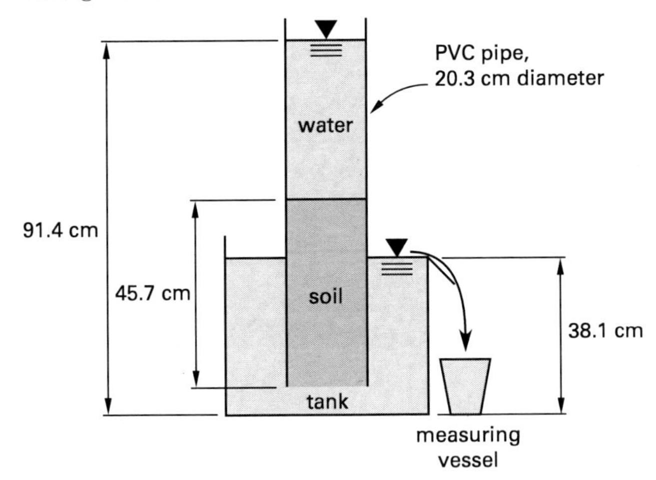

The coefficient of permeability is most nearly

(A) 4.2×10^{-5} cm/s

(B) 5.7×10^{-5} cm/s

(C) 2.5×10^{-3} cm/s

(D) 6.0×10^{-1} cm/s

Hint: Determine the coefficient of permeability by using Darcy's law.

PROBLEM 14

A falling-head permeability test with the parameters shown was performed on a laboratory sample. According to the test results, it takes 75 s for water to stop flowing through the exit valve.

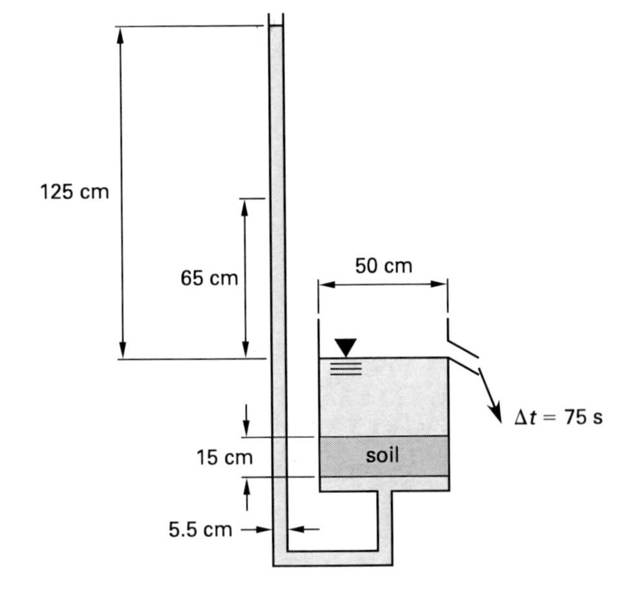

The coefficient of permeability is most nearly

(A) 1.2×10^{-3} cm/s

(B) 1.6×10^{-3} cm/s

(C) 1.2 cm/s

(D) 11 cm/s

Hint: The equation for the falling-head permeability calculation involves a log expression.

PROBLEM 15

Steady-state horizontal seepage is occurring in the cylindrical permeability test sample shown. Prior to commencement of the test, the weight of the empty apparatus was 11.71 lbf. After placement of the dry soil sample, the apparatus weighed 15.29 lbf.

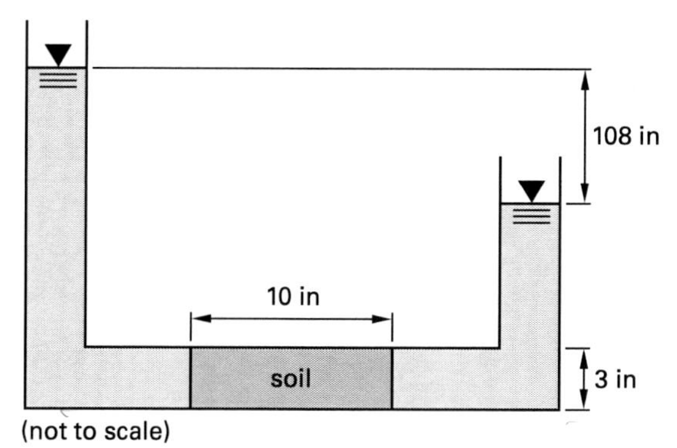

Approximate the seepage velocity given that the permeability of the sample is 1.75×10^{-2} in/sec and the specific gravity is 2.65.

(A) 0.089 in/sec

(B) 0.19 in/sec

(C) 0.21 in/sec

(D) 0.40 in/sec

Hint: The seepage velocity is greater than the discharge velocity and is calculated using the porosity.

STRENGTH TESTING OF SOIL AND ROCK

PROBLEM 16

A triaxial test was performed on a cylindrical silty sand specimen. The specimen was sheared undrained under confinement. The conditions at failure are shown.

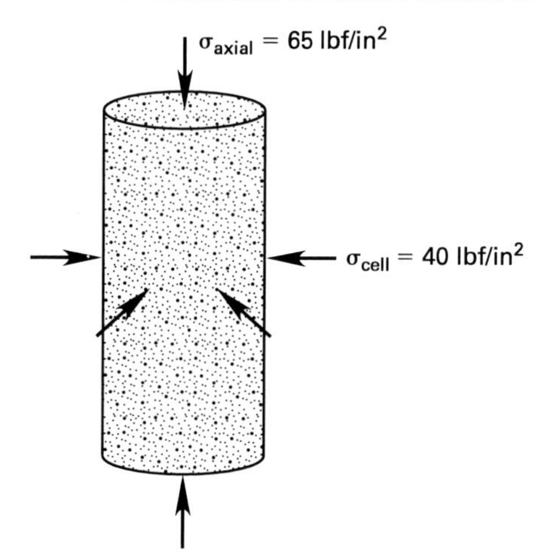

Given that the pore pressure at failure was 20 lbf/in^2 and assuming the specimen has negligible cohesion, the effective angle of internal friction is most nearly

(A) 4.0°

(B) 14°

(C) 23°

(D) 32°

Hint: Use the obliquity relationship to determine the effective angle of internal friction.

STRESS-STRAIN TESTING OF SOIL AND ROCK

PROBLEM 17

For soil with the given characteristics, what is the expected shear strength at the midpoint of the silty sand layer?

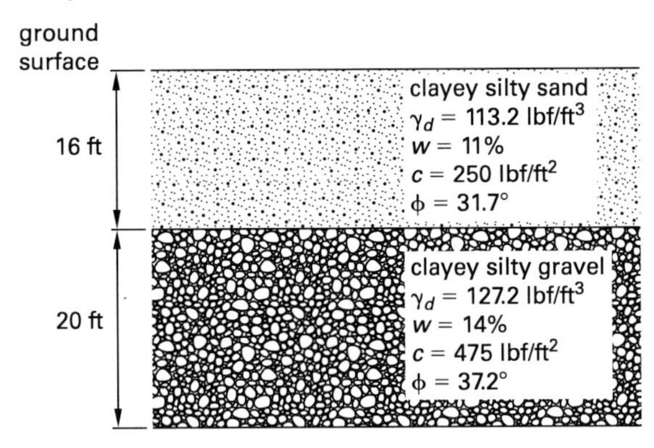

(A) 155 lbf/ft^2

(B) 250 lbf/ft^2

(C) 795 lbf/ft^2

(D) 870 lbf/ft^2

Hint: Use the given unit weight to calculate the confining pressure due to overburden.

PROBLEM 18

A 5 ft thick clay layer commences at a depth of approximately 8 ft below the ground surface and is overlain by a relatively porous sandy soil and underlain by gravel and cobbles. A consolidation test was performed on a sample of the clay with a resulting coefficient of vertical consolidation of 7.3×10^{-7} ft^2/sec. The approximate number of days required to achieve 90% consolidation by preloading with a large fill is most nearly

(A) 20 days

(B) 80 days

(C) 85 days

(D) 340 days

Hint: The length of time is dependent on pore pressure dissipation in the clay layer.

PROBLEM 19

Using the soil profile data given, what is most nearly the time required for 75% consolidation?

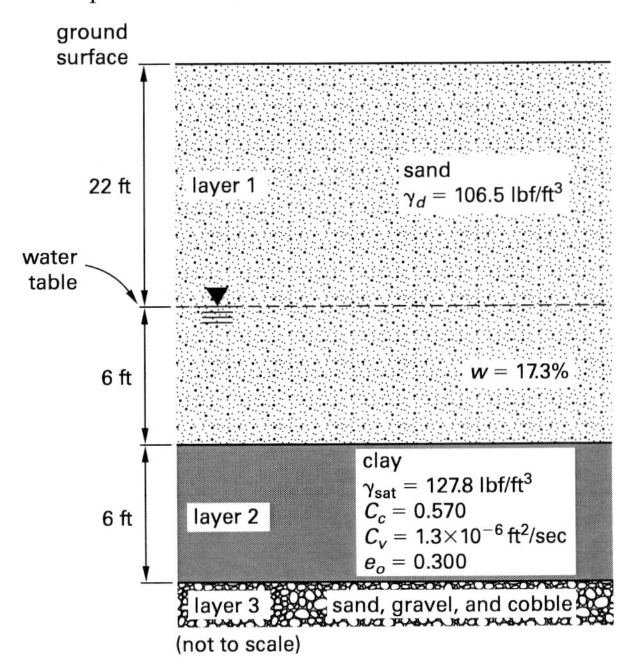

(A) 15 days

(B) 40 days

(C) 70 days

(D) 160 days

Hint: The quantity of time is independent of the initial effective stress or the change in total stress. Use the time rate of consolidation relationship derived from the Terzaghi one-dimensional consolidation equation.

PROBLEM 20

Approximate the time required for 80% consolidation of the clay layer shown if the coefficient of vertical consolidation is 4.3×10^{-6} ft^2/sec.

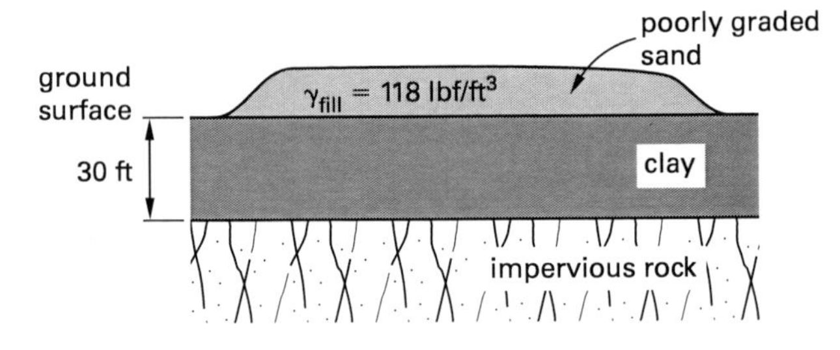

(A) 25 days

(B) 350 days

(C) 1400 days

(D) 2100 days

Hint: The quantity of time depends on the length of the drainage path.

PROBLEM 21

A field technician submits the results of a field density test for review. Prior to the test, the laboratory determined the unit weight of the dry test sand to be 84.7 lbf/ft^3. The soil obtained from the test hole was weighed at 3.65 lbf and had a moisture content at 17.3%. The weight of the sand that filled the test hole was 2.23 lbf. Assuming the field moisture content is within the acceptable specification range for optimum moisture content, what is the in-place percent compaction if the laboratory Proctor test results in a maximum dry density of 122.3 lbf/ft^3?

(A) 97%

(B) 104%

(C) 114%

(D) 133%

Hint: Determine the volume of the test hole by using the unit weight properties of the test sand.

SOLUTION 1

The relationship between the total stress, σ, effective stress, σ', and pore pressure, u, can be expressed as

$$\sigma_{\text{ave}} = \sigma' + u$$

Given the unit weight of each soil, the layer thickness, and the location of the groundwater table, the total stress can be determined at the midpoint of the sandy clay layer. The total stress value for layer A is first determined, and then the total stress for each layer (or partial layer as necessary) is subsequently added to determine the desired total stress at the midpoint of the sandy clay layer.

Layer B should be reduced to sublayers for treatment of the different water content values.

$$\sigma_{\text{ave}} = \sigma_{\text{A}} + \Delta\sigma_{\text{B}_1} + \Delta\sigma_{\text{B}_2} + \Delta\sigma_{\text{C,midpoint}}$$

Only a partial increase in the total stress for layer C is required to be determined.

Calculate the total pressure at the base of layer A.

$$\begin{aligned} \sigma_{\text{A}} &= \gamma_{\text{A}} D_{\text{A}} \\ &= \left(112.3\ \frac{\text{lbf}}{\text{ft}^3}\right)(9\ \text{ft}) \\ &= 1011\ \text{lbf/ft}^2 \end{aligned}$$

Calculate the additional total stress both at the base of layer B and midway through layer C, keeping in mind the conversion from dry unit weight to moist (total) unit weight where necessary.

$$\gamma_t = \gamma_d(1 + w)$$

Break the calculation for layer B into two parts for treatment of different water content values.

For the top portion of layer B (above the phreatic surface), layer B_1,

$$\begin{aligned} \Delta\sigma_{\text{B}_1} &= \gamma_t D = \gamma_{d,\text{B}}(1 + w_{\text{B}_1}) D_{\text{B}_1} \\ &= \left(108.2\ \frac{\text{lbf}}{\text{ft}^3}\right)(1 + 0.072)(14\ \text{ft}) \\ &= 1624\ \text{lbf/ft}^2 \end{aligned}$$

For the bottom portion of layer B (below the phreatic surface), layer B_2,

$$\begin{aligned}\Delta\sigma_{B_2} &= \gamma_t D = \gamma_{d,B}(1+w_{B_2})D_{B_2}\\ &= \left(108.2\ \frac{\text{lbf}}{\text{ft}^3}\right)(1+0.146)(18\ \text{ft})\\ &= 2232\ \text{lbf/ft}^2\end{aligned}$$

Calculate the additional total stress at the midpoint in layer C.

$$\begin{aligned}\Delta\sigma_{C,\text{midpoint}} &= \gamma_t D = \gamma_{d,C}(1+w_C)D_{C,\text{midpoint}}\\ &= \left(111.7\ \frac{\text{lbf}}{\text{ft}^3}\right)(1+0.198)(6.5\ \text{ft})\\ &= 870\ \text{lbf/ft}^2\end{aligned}$$

Determine the total stress at the midpoint of the sandy clay layer by adding all the pressure values previously determined.

$$\begin{aligned}\sigma_{\text{ave}} &= \sigma_A + \Delta\sigma_{B_1} + \Delta\sigma_{B_2} + \Delta\sigma_{C,\text{midpoint}}\\ &= 1011\ \frac{\text{lbf}}{\text{ft}^2} + 1624\ \frac{\text{lbf}}{\text{ft}^2} + 2232\ \frac{\text{lbf}}{\text{ft}^2} + 870\ \frac{\text{lbf}}{\text{ft}^2}\\ &= 5737\ \text{lbf/ft}^2\end{aligned}$$

The pore pressure at the midpoint of the clay layer is determined using the unit weight of water in reference to the depth of the groundwater phreatic surface.

$$\begin{aligned}u &= \gamma_w D_{\text{phreatic}}\\ &= \left(62.4\ \frac{\text{lbf}}{\text{ft}^3}\right)(18\ \text{ft} + 6.5\ \text{ft})\\ &= 1529\ \text{lbf/ft}^2\end{aligned}$$

The effective stress at the midpoint of the clay layer can be calculated by subtracting the pore pressure from the total stress.

$$\begin{aligned}\sigma' &= \sigma_{\text{ave}} - u\\ &= 5737\ \frac{\text{lbf}}{\text{ft}^2} - 1529\ \frac{\text{lbf}}{\text{ft}^2}\\ &= 4208\ \text{lbf/ft}^2\quad (4210\ \text{lbf/ft}^2)\end{aligned}$$

The answer is (B).

Why Other Options Are Wrong

(A) This incorrect solution is obtained by using the dry unit weight for each layer rather than the moist (total) unit weight where required by moisture conditions and groundwater conditions. The result is a low value compared to the correct answer.

(C) This incorrect solution results from mistakenly finding the effective stress at the bottom of the clay layer, not at the midpoint. A typical error is to assume that the effective stress at the base of the clay layer will be comparable to the midpoint value, which may be acceptable if the layer is relatively thin. However, common practice is to use the midpoint where possible if the midpoint properties can be determined.

(D) This incorrect solution is the total stress at the midpoint of the clay layer and is obtained by failing to subtract the pore water pressure from the total stress, a common oversight.

SOLUTION 2

The porosity is given for the sand layer (layer 1). Determine the void ratio and use it to determine the dry unit weight and saturated unit weight of the sand.

$$e_1 = \frac{n}{1-n} = \frac{0.5}{1-0.5} = 1$$

The dry density can be expressed in terms of the specific gravity, void ratio, and unit weight of water. However, the specific gravity and weight of water terms can be replaced by the unit weight of solids.

$$\begin{aligned}\gamma_{d,1} &= \frac{(\text{SG})\gamma_w}{1+e_1} = \frac{\gamma_{s,1}}{1+e_1} = \frac{170.6\ \frac{\text{lbf}}{\text{ft}^3}}{1+1}\\ &= 85.3\ \text{lbf/ft}^3\end{aligned}$$

The degree of saturation for the sand can be expressed using the following relationship.

$$S = \frac{w(\text{SG})}{e_1} = \frac{w\left(\frac{\gamma_{s,1}}{\gamma_w}\right)}{e_1}$$

The sand below the water table can be assumed to be saturated. The expression can be solved for the saturated water content.

$$\begin{aligned}w_{\text{sat}} &= \frac{S\gamma_w e_1}{\gamma_{s,1}} = \frac{(1)\left(62.4\ \frac{\text{lbf}}{\text{ft}^3}\right)(1)}{170.6\ \frac{\text{lbf}}{\text{ft}^3}}\\ &= 0.366\end{aligned}$$

The saturated unit weight for the sand can be determined.

$$\begin{aligned}\gamma_{\text{sat},1} &= \gamma_{d,1}(1+w_{\text{sat}}) \\ &= \left(85.3\ \frac{\text{lbf}}{\text{ft}^3}\right)(1+0.366) \\ &= 116.5\ \text{lbf/ft}^3\end{aligned}$$

Calculate the effective stress at the base of the clay layer.

$$\begin{aligned}\sigma' &= \gamma_{d,1}t_{1,\text{above}} + (\gamma_{\text{sat},1}-\gamma_w)t_{1,\text{below}} + (\gamma_{\text{sat},2}-\gamma_w)t_2 \\ &= \left(85.3\ \frac{\text{lbf}}{\text{ft}^3}\right)(12\ \text{ft}) + \left(116.5\ \frac{\text{lbf}}{\text{ft}^3} - 62.4\ \frac{\text{lbf}}{\text{ft}^3}\right) \\ &\quad \times(14\ \text{ft}) + \left(127.8\ \frac{\text{lbf}}{\text{ft}^3} - 62.4\ \frac{\text{lbf}}{\text{ft}^3}\right)(22\ \text{ft}) \\ &= 3220\ \text{lbf/ft}^2\end{aligned}$$

The answer is (A).

Why Other Options Are Wrong

(B) This incorrect solution assumes the porosity value is the voids ratio. It uses the void ratio equation with a unit value for the volume of solids in the phase relationship calculations to produce incorrect dry and saturated unit weights for the sand layer. The result is a slightly higher value compared to the correct answer.

(C) This incorrect solution is obtained by using the solids unit weight for the sand layer as the dry unit weight and then proceeding to use that value to calculate the saturated unit weight for the sand layer and the subsequent effective stress at the base of the clay layer. The result is a high value compared to the correct answer.

(D) This incorrect solution can be obtained by calculating the total stress at the bottom of the clay layer without subtracting the pore pressure.

SOLUTION 3

Since groundwater (hydrostatic conditions) is not present, the effective stress and the total stress are equal. The coefficient of lateral earth pressure at rest can be used to express the ratio of the horizontal stress to the vertical stress.

$$k_o = \frac{\sigma_h}{\sigma_v}$$

The horizontal stress can be calculated as

$$\begin{aligned}\sigma_h &= k_o\sigma_v \\ &= k_o\gamma_m D \\ &= (0.4)\left(115\ \frac{\text{lbf}}{\text{ft}^3}\right)(23\ \text{ft}) \\ &= 1058\ \text{lbf/ft}^2 \quad (1060\ \text{lbf/ft}^2)\end{aligned}$$

The answer is (B).

Why Other Options Are Wrong

(A) This incorrect solution converts the given moist unit weight to a dry unit weight prior to performing the calculation of horizontal stress.

(C) This incorrect solution multiplies the given moist unit weight by the given water content to obtain an incorrect higher moist unit weight value prior to determining the horizontal stress.

(D) This incorrect solution is the horizontal stress at the base of the given soil layer at 29 ft instead of at the requested 23 ft below the ground surface.

SOLUTION 4

The plasticity index, PI, the Hazen uniformity coefficient, C_u, and the coefficient of curvature, C_z, should be calculated to assist with the soil classification procedure.

$$\begin{aligned}\text{PI} &= \text{LL} - \text{PL} \\ C_u &= \frac{D_{60}}{D_{10}} \\ C_z &= \frac{(D_{30})^2}{D_{10}D_{60}}\end{aligned}$$

Calculate the PI.

$$\text{PI} = \text{LL} - \text{PL} = 31 - 25 = 6$$

Assign the grain diameter coefficients.

$$\begin{aligned}D_{10} &= \text{no. 200 sieve opening} = 0.075\ \text{mm} \\ D_{30} &= \text{no. 40 sieve opening} = 0.425\ \text{mm} \\ D_{60} &= \text{no. 10 sieve opening} = 2.00\ \text{mm}\end{aligned}$$

Calculate the Hazen uniformity coefficient.

$$C_u = \frac{D_{60}}{D_{10}} = \frac{2.00\ \text{mm}}{0.075\ \text{mm}} = 27$$

Calculate the coefficient of curvature.

$$C_z = \frac{(D_{30})^2}{D_{10}D_{60}} = \frac{(0.425\ \text{mm})^2}{(0.075\ \text{mm})(2.00\ \text{mm})} = 1.20$$

Use the USCS chart to determine the major division between coarse- and fine-grained soils using the sieve analysis results. Because 10% of the total weight is finer than the no. 200 sieve, 90% is retained on or above this sieve, indicating a coarse-grained soil. Also, $(20\%)/(90\%) \times 100 = 22.2\%$ of the coarse fraction is retained on the no. 4 sieve, leaving $100\% - 22.2\% = 77.8\%$ of the coarse fraction passing the no. 4 sieve. Therefore, the criteria for sand being over half of the coarse fraction is met.

In the USCS chart, there is a column labeled "laboratory classification criteria," with a sub-column labeled "supplementary criteria requirements," where C_u and C_z are listed. The values of C_u and C_z were previously calculated to be 27 and 1.20, respectively. The criteria indicate that the classification symbol must start with an SW instead of one of the remaining choices (SP, SM, or SC), because only the supplementary requirements for SW are met for this soil.

Since the percentage passing the no. 200 sieve is within the range of 5–12%, the USCS chart indicates that the result is a borderline case requiring the use of a "dual symbol." The Atterberg limits, LL = 31 and PI = 6, both plot on the plasticity chart below the "A-line" in the ML zone, indicating that the second part of the dual symbol is SM, making the correct answer SW-SM.

The answer is (D).

Why Other Options Are Wrong

(A) A classification of SM is possible because both the Atterberg limits plot on the plasticity chart below the "A-line." However, this answer fails to meet the criteria for a single symbol because the soil has less than 12% passing the no. 200 sieve.

(B) This solution is incorrect because the percentage passing the no. 200 sieve is greater than five. Therefore, the criteria for SP are not met.

(C) Although the result for C_u is greater than six and C_z is between one and three, LL and PI are plotted on the plasticity chart in the ML zone rather than in the CL zone. Therefore, a symbol of SC cannot be used as part of the required dual symbol.

SOLUTION 5

The plasticity index, PI, can be calculated from the liquid limit, LL, and the plastic limit, PL.

$$\begin{aligned} \text{PI} &= \text{LL} - \text{PL} = 29 - 19 \\ &= 10 \end{aligned}$$

Use the AASHTO soil classification flow chart from left to right to determine the soil group. Check the criteria, and move to the next column (immediate right) if any of the criteria fail in the first column.

Laboratory data for the percentage passing the no. 10 or no. 40 sieve sizes is not provided because the percentage passing the no. 8 sieve is 100%. Therefore, the percent passing the no. 200 sieve criteria is the first sieve analysis requirement to pass, indicating the soil type is in the A-4 group. The remaining criteria for liquid limit and plasticity index are also met. Therefore, B1 (5–6 ft) is an A-4 group classification.

The answer is (B).

Why Other Options Are Wrong

(A) This solution mistakenly applies the no. 200 sieve criteria and the Atterberg limits criteria. A-3 classification requires the soil to be non-plastic.

(C) This solution is obtained by applying the Atterberg limits criteria (specifically the plasticity index) incorrectly. A-6 classification requires the plasticity index be a minimum of 11.

(D) This solution is obtained by applying the liquid limit criteria incorrectly. Specifically, for soil classification A-7-6, the minimum liquid limit is required to be 41.

SOLUTION 6

The plasticity index, PI, can be calculated from the liquid limit, LL, and the plastic limit, PL.

$$\begin{aligned} \text{PI} &= \text{LL} - \text{PL} = 38 - 20 \\ &= 18 \end{aligned}$$

Use the USCS chart to determine the major division between coarse- and fine-grained soils using the sieve analysis results. Because 72% of the total weight is finer than the no. 200 sieve, the soil is a fine-grained soil. The classification symbol must start with a "C," "M," or "O." However, "O" is not given as an answer choice. The PI and LL are 18 and 38, respectively, and both plot on the plasticity chart above the "A-line" in the CL zone, indicating the symbol is CL.

The answer is (A).

Why Other Options Are Wrong

(B) This solution mistakenly plots the plasticity information on the plasticity chart in the CH region.

(C) This solution mistakenly plots the plasticity information on the plasticity chart in the ML region.

(D) This solution mistakenly plots the plasticity information on the plasticity chart in the MH region.

SOLUTION 7

The PL of the soil is the water content of the sample as a percentage, but presented without an actual percentage symbol. The water content is calculated as the quantity or mass of water contained in the sample divided by the dry mass of the sample.

$$w = \frac{m_w}{m_s} \times 100\%$$

The mass of water can be determined by subtracting the dry mass plus container value from the wet mass plus container value, leaving only the mass of water.

$$\begin{aligned} m_w &= m_{\text{wsc}} - m_{\text{dsc}} \\ &= 23.42 \text{ g} - 19.81 \text{ g} \\ &= 3.61 \text{ g} \end{aligned}$$

The dry mass of the sample can be determined by subtracting the mass of the container itself from the dry mass plus container value.

$$\begin{aligned} m_s &= m_{\text{dsc}} - m_c \\ &= 19.81 \text{ g} - 1.73 \text{ g} \\ &= 18.08 \text{ g} \end{aligned}$$

The water content (expressed as a percentage) is determined by dividing the mass of water contained in the sample by the dry weight of the sample.

$$\begin{aligned} w &= \frac{m_w}{m_s} \times 100\% \\ &= \frac{3.61 \text{ g}}{18.08 \text{ g}} \times 100\% \\ &= 20\% \end{aligned}$$

The PL is a percentage value corresponding to the water content and is commonly expressed without the percentage symbol.

$$\text{PL} = 20$$

The answer is (B).

Why Other Options Are Wrong

(A) This incorrect solution is obtained by dividing the mass of water by the dry mass of soil plus container value without first subtracting the mass of the container itself. The value is less than the correct answer.

(C) This solution is incorrect because, although this value uses the correct mass of solids, the mass of solids is subtracted from the wet mass plus container value without first subtracting the mass of the container itself, resulting in too large a value for the mass of water. That value is then divided by the mass of solids, resulting in a value greater than the correct answer.

(D) This incorrect solution divides the dry mass plus container value by the wet mass plus container value, obtaining a result that is much greater than the correct answer.

SOLUTION 8

The void ratio can be expressed as

$$\begin{aligned} e &= \frac{(\text{SG})V_t\gamma_w}{M_s} - 1 \\ &= \frac{(2.66)(0.44 \text{ ft}^3)\left(62.4 \ \frac{\text{lbf}}{\text{ft}^3}\right)}{43 \text{ lbf}} - 1 \\ &= 0.7 \end{aligned}$$

The answer is (C).

Why Other Options Are Wrong

(A) This incorrect answer uses the total weight of the sample instead of the dry weight of the sample, producing a result that is less than the correct answer.

(B) This incorrect solution is actually the porosity determined by a simplified phase relationship equation (not including the percentage symbol). The result is less than the correct answer.

(D) Using the total volume divided by the volume of the solids to calculate the void ratio produces this incorrect answer. The value is greater than the correct answer and is the same as using the simplified equation and not subtracting 1.

SOLUTION 9

The degree of saturation can be expressed in terms of water content, void ratio, and specific gravity.

$$S = \frac{w_{\text{sat}}(\text{SG})}{e}$$

The expression can then be solved in terms of the saturated water content.

$$w_{\text{sat}} = \frac{Se}{\text{SG}}$$

The dry weight and total volume are both given in the problem statement. The dry density can be calculated.

$$\gamma_d = \frac{W_s}{V_t} = \frac{43 \text{ lbf}}{0.39 \text{ ft}^3} = 110.3 \text{ lbf/ft}^3$$

The dry density can also be expressed in terms of the specific gravity, void ratio, and unit weight of water. Then, the expression for dry density can be solved for the void ratio.

$$\gamma_d = \frac{(\text{SG})\gamma_w}{1+e}$$

$$e = \frac{(\text{SG})\gamma_w}{\gamma_d} - 1 = \frac{(2.66)\left(62.4 \ \frac{\text{lbf}}{\text{ft}^3}\right)}{110.3 \ \frac{\text{lbf}}{\text{ft}^3}} - 1 = 0.50$$

The saturated water content can be calculated (and expressed as a percentage).

$$w_{\text{sat}} = \frac{Se}{\text{SG}} = \frac{(1.0)(0.50)}{2.66} = 0.19 \quad (19\%)$$

The answer is (B).

Why Other Options Are Wrong

(A) This incorrect solution is obtained by using the total weight of the sample instead of the weight of the solids.

(C) This incorrect solution is actually the water content of the soil.

(D) This incorrect solution is obtained by using the volume of the solids instead of the volume of the voids to find the weight of water and then proceeding to find the maximum water content.

SOLUTION 10

The sample recovery was made using five rings. The total mass of the soil sample can be calculated as the total mass of field sample minus the mass of five rings.

$$\begin{aligned} m_{s,\text{total}} &= m_{\text{fs,total}} - 5m_{\text{ring}} \\ &= 879 \text{ g} - (5)(44 \text{ g}) \\ &= 659 \text{ g} \end{aligned}$$

The total density of the recovered soil sample can be calculated and converted to dry unit weight.

$$\begin{aligned} V_{\text{rings,total}} &= (\text{no. of rings})\left(\tfrac{1}{4}\pi D^2\right)H \\ &= (5)\left(\frac{1}{4}\right)\pi\left((2.42 \text{ in})\left(2.54 \ \frac{\text{cm}}{\text{in}}\right)\right)^2 \\ &\quad \times (1 \text{ in})\left(2.54 \ \frac{\text{cm}}{\text{in}}\right) \\ &= 377 \text{ cm}^3 \end{aligned}$$

$$\begin{aligned} \gamma_t &= \frac{W_s}{5V_{\text{sr}}} = \left(\frac{m_{s,\text{total}}}{V_{\text{rings,total}}}\right)\gamma_w \\ &= \left(\frac{659 \text{ g}}{377 \text{ cm}^3}\right)\left(\frac{62.4 \ \frac{\text{lbf}}{\text{ft}^3}}{1 \ \frac{\text{g}}{\text{cm}^3}}\right) \\ &= 109.1 \text{ lbf/ft}^3 \end{aligned}$$

$$\begin{aligned} \gamma_d &= \frac{\gamma_t}{1+w} = \frac{109.1 \ \frac{\text{lbf}}{\text{ft}^3}}{1+0.112} \\ &= 98 \text{ lbf/ft}^3 \quad (100 \text{ lbf/ft}^3) \end{aligned}$$

The answer is (A).

Why Other Options Are Wrong

(B) This incorrect solution is the total unit weight, obtained by failing to convert to dry unit weight.

(C) This incorrect solution is obtained by subtracting the mass of one ring instead of five rings in determining the total mass of the soil sample. This dry unit weight value is high compared to the correct answer.

(D) This incorrect solution fails to subtract the mass of five rings before determining the total unit weight. This dry unit weight value is very high compared to the correct answer.

SOLUTION 11

The void ratio is defined as the volume of the voids divided by the volume of the solids and is typically not expressed as a percentage. For the given saturated conditions and assuming a unit value for the volume of the solids, a simplified method based on the phase relationship between the degree of saturation and the void ratio can be used.

$$S = \frac{w(\text{SG})}{e} = \frac{w\left(\frac{\gamma_s}{\gamma_w}\right)}{e}$$

Solve the relationship for the void ratio, knowing that the degree of saturation is 1.

$$e = \frac{w\left(\frac{\gamma_s}{\gamma_w}\right)}{S} = \frac{(0.18)\left(\frac{170\ \frac{\text{lbf}}{\text{ft}^3}}{62.4\ \frac{\text{lbf}}{\text{ft}^3}}\right)}{1} = 0.49$$

The answer is (B).

Why Other Options Are Wrong

(A) This solution is the porosity rather than the requested void ratio. This mistake is commonly made by erroneously determining the porosity instead of the void ratio while still assuming a unit value for the volume of solids.

(C) This solution fails to first multiply the weight of the solids by the water content to find the weight of the water. It then proceeds to divide the incorrect weight of water value by the unit weight of water. The units do not work out, and the value is greater than the correct answer.

(D) This solution is the reciprocal of the correct solution and is obtained by dividing the volume of solids by the volume of voids. The value is greater than the correct answer.

SOLUTION 12

The porosity can be defined using the void ratio and is usually expressed as a percentage.

$$n = \frac{e}{e+1} \times 100\%$$

By assuming a unit value for the total volume, an equation for the void ratio can be simplified to determine the porosity.

$$e = \frac{V_t\gamma_w(\text{SG})}{W_s} - 1 = \frac{V_t\gamma_w\left(\frac{\gamma_s}{\gamma_w}\right)}{W_s} - 1$$
$$= \frac{\gamma_s}{\gamma_d} - 1$$

The dry unit weight can be determined based on the total unit weight and the moisture content.

$$\gamma_d = \frac{\gamma_t}{1+w} = \frac{127.0\ \frac{\text{lbf}}{\text{ft}^3}}{1+0.1}$$
$$= 115.5\ \text{lbf/ft}^3$$

The void ratio can be determined.

$$e = \frac{\gamma_s}{\gamma_d} - 1 = \frac{165.4\ \frac{\text{lbf}}{\text{ft}^3}}{115.5\ \frac{\text{lbf}}{\text{ft}^3}} - 1$$
$$= 0.43$$

The porosity can be determined.

$$n = \frac{e}{e+1} \times 100\% = \frac{0.43}{0.43+1} \times 100\% = 30\%$$

The answer is (D).

Why Other Options Are Wrong

(A) This incorrect solution is actually the void ratio expressed as a percentage.

(B) This incorrect solution is the reciprocal of the porosity without converting the number to a percentage. A common error is caused by a failure to understand the equation and its reciprocal.

(C) This incorrect solution makes the common mistake of dividing the weight of solids by the total unit weight rather than by the solids unit weight to determine the volume of solids. This results in a value that is greater than the correct value.

SOLUTION 13

Use Darcy's law for the constant-head test.

$$Q = KiA$$

Solve for coefficient of permeability.

$$K = \frac{Q}{iA}$$

Determine the test sample cross-sectional area.

$$A = \frac{\pi d^2}{4} = \frac{\pi(20.3\ \text{cm})^2}{4} = 324\ \text{cm}^2$$

The hydraulic gradient is the change in head through the sample divided by the length of the soil sample. The hydraulic gradient is dimensionless.

$$i = \frac{\Delta h}{L}$$
$$= \frac{91.4\ \text{cm} - 38.1\ \text{cm}}{45.7\ \text{cm}}$$
$$= 1.17$$

During the test duration of 4 h, 227 g of water are collected. Therefore, Q can be calculated from the mass and the equivalent volume of water collected during the 4 h test.

$$Q = \frac{V}{t} = \frac{\frac{m_w}{\rho_w}}{t} = \frac{\frac{227\ \text{g}}{1\ \frac{\text{g}}{\text{cm}^3}}}{(4\ \text{h})\left(3600\ \frac{\text{s}}{\text{h}}\right)} = 1.58 \times 10^{-2}\ \text{cm}^3/\text{s}$$

The coefficient of permeability can be calculated.

$$K = \frac{Q}{iA} = \frac{1.58 \times 10^{-2}\ \frac{\text{cm}^3}{\text{s}}}{(1.17)(324\ \text{cm}^2)} = 4.2 \times 10^{-5}\ \text{cm/s}$$

The answer is (A).

Why Other Options Are Wrong

(B) This incorrect solution uses the reciprocal of the hydraulic gradient during the calculation. This is a common mistake based on erroneous interpretation of the problem setup and misunderstanding how the hydraulic gradient is determined. The result is greater than the correct answer.

(C) This incorrect solution is obtained when the volume of water collected is divided by 240 min to obtain the wrong value for the volumetric flow rate. A common mistake is to only multiply the time in hours by 60 to get 240 s (actually minutes) and use that in the calculation rather than 3600 s/h to get 14 400 s. The result is greater than the correct answer.

(D) This incorrect solution, a common oversight, is obtained when the volume of water collected is not divided by a time value to obtain a volumetric flow rate. The result is much greater than the correct answer, and the units do not work out.

SOLUTION 14

The equation for calculating the coefficient of permeability using a falling-head apparatus is derived from Darcy's law.

$$K = 2.3\left(\frac{aL}{A\Delta t}\right)\log_{10}\frac{h_1}{h_2}$$

Calculate the areas of the standpipe, a, and soil sample, A.

$$a = \tfrac{1}{4}\pi d^2 = \tfrac{1}{4}\pi(5.5\ \text{cm})^2 = 23.8\ \text{cm}^2$$

$$A = \tfrac{1}{4}\pi D^2 = \tfrac{1}{4}\pi(50\ \text{cm})^2 = 1963.5\ \text{cm}^2$$

The coefficient of permeability can be calculated.

$$K = 2.3\left(\frac{aL}{A\Delta t}\right)\log_{10}\frac{h_1}{h_2} = (2.3)\left(\frac{(23.8\ \text{cm}^2)(15\ \text{cm})}{(1963.5\ \text{cm}^2)(75\ \text{s})}\right)\log_{10}\frac{125\ \text{cm}}{65\ \text{cm}} = 1.58 \times 10^{-3}\ \text{cm/s} \quad (1.6 \times 10^{-3}\ \text{cm/s})$$

The answer is (B).

Why Other Options Are Wrong

(A) This solution erroneously interprets the falling-head permeability equation as a constant-head permeability equation by finding a volumetric flow rate and hydraulic gradient. The value is less than the correct answer.

(C) This solution fails to divide by the change in time. The value is greater than the correct value and the units are not correct.

(D) This solution erroneously interchanges the standpipe area and soil specimen area in the falling-head permeability equation. The value is greater than the correct answer.

SOLUTION 15

The porosity of the soil sample is required to calculate the seepage velocity.

$$n = \frac{e}{e+1}$$

The dry density can be expressed in terms of the specific gravity, void ratio, and unit weight of water.

$$\gamma_d = \frac{(\text{SG})\gamma_w}{1+e}$$

The expression for dry density can be solved for the void ratio.

$$e = \frac{(\text{SG})\gamma_w}{\gamma_d} - 1$$

The dry density can be calculated from the change in weight of the sample apparatus.

$$\begin{aligned}\gamma_d &= \frac{W_s}{V_t} = \frac{\Delta W_{\text{sample apparatus}}}{AL} \\ &= \frac{W_f - W_i}{\left(\frac{\pi D^2}{4}\right)L} \\ &= \frac{(15.29 \text{ lbf} - 11.71 \text{ lbf})\left(12 \frac{\text{in}}{\text{ft}}\right)^3}{\left(\frac{\pi(3 \text{ in})^2}{4}\right)(10 \text{ in})} \\ &= 87.5 \text{ lbf/ft}^3\end{aligned}$$

The void ratio can be calculated.

$$\begin{aligned}e &= \frac{(\text{SG})\gamma_w}{\gamma_d} - 1 = \frac{(2.65)\left(62.4 \frac{\text{lbf}}{\text{ft}^3}\right)}{87.5 \frac{\text{lbf}}{\text{ft}^3}} - 1 \\ &= 0.89\end{aligned}$$

Darcy's law can be used to calculate the discharge velocity. The flow rate is not necessary for the calculation because the hydraulic gradient is known.

$$\begin{aligned}\text{v}_d &= Ki = K\left(\frac{\Delta h}{L}\right) \\ &= \left(1.75 \times 10^{-2} \frac{\text{in}}{\text{sec}}\right)\left(\frac{108 \text{ in}}{10 \text{ in}}\right) \\ &= 0.19 \text{ in/sec}\end{aligned}$$

The discharge velocity is related to the seepage velocity by the porosity and can be calculated.

$$\text{v}_d = n\text{v}_s$$

The porosity can be calculated.

$$n = \frac{e}{e+1} = \frac{0.89}{0.89+1} = 0.47$$

The seepage velocity is

$$\begin{aligned}\text{v}_s &= \frac{\text{v}_d}{n} = \frac{0.19 \frac{\text{in}}{\text{sec}}}{0.47} \\ &= 0.40 \text{ in/sec}\end{aligned}$$

The answer is (D).

Why Other Options Are Wrong

(A) This incorrect solution multiplies the discharge velocity and the porosity rather than dividing by the porosity. The result is less than the correct result.

(B) This incorrect solution is actually the discharge velocity. The result is less than the correct answer.

(C) This incorrect solution mistakenly uses the void ratio instead of the porosity to determine the seepage velocity. The result is low compared to the correct answer.

SOLUTION 16

By assuming negligible cohesion, the effective angle of internal friction may be determined using the obliquity relationship.

$$\sin\phi' = \frac{\sigma'_{1f} - \sigma'_{3f}}{\sigma'_{1f} + \sigma'_{3f}}$$

Calculate the effective principal stresses.

$$\begin{aligned}\sigma'_{1f} &= \sigma_{\text{axial}} - u = 65 \frac{\text{lbf}}{\text{in}^2} - 20 \frac{\text{lbf}}{\text{in}^2} \\ &= 45 \text{ lbf/in}^2\end{aligned}$$

$$\begin{aligned}\sigma'_{3f} &= \sigma_{\text{cell}} - u = 40 \frac{\text{lbf}}{\text{in}^2} - 20 \frac{\text{lbf}}{\text{in}^2} \\ &= 20 \text{ lbf/in}^2\end{aligned}$$

Determine the effective angle of internal friction.

$$\begin{aligned}\phi' &= \arcsin\frac{\sigma'_{1f} - \sigma'_{3f}}{\sigma'_{1f} + \sigma'_{3f}} \\ &= \arcsin\frac{45 \frac{\text{lbf}}{\text{in}^2} - 20 \frac{\text{lbf}}{\text{in}^2}}{45 \frac{\text{lbf}}{\text{in}^2} + 20 \frac{\text{lbf}}{\text{in}^2}} \\ &= 23°\end{aligned}$$

The answer is (C).

Why Other Options Are Wrong

(A) This incorrect answer is obtained if the total confining stress is subtracted from the effective major principal stress at failure. The pore pressure is subtracted from the axial stress at failure only and not from the confining pressure before using the obliquity relationship. Subtracting the total confining pressure from the effective major principal stress results in a very low value compared to the correct value for effective angle of internal friction.

(B) This incorrect solution is the total angle of internal friction and is obtained by not subtracting pore pressure from the failure conditions to determine the effective principal stress difference. The value for effective angle of internal friction is larger than the total angle of internal friction.

(D) Subtracting the effective confining stress from the total major principal stress at failure results in this incorrect solution. The pore pressure was subtracted from the confining stress at failure and not from the total major principal stress before using the obliquity relationship. Subtracting the effective confining pressure from the total principal stress results in a very high value compared to the correct value for effective angle of internal friction.

SOLUTION 17

Use the equation for the shear strength in terms of total stress.

$$S = c + \sigma \tan\phi$$

Determine the overburden (confining) pressure at the midpoint of the clayey silty sand layer.

$$\begin{aligned}\sigma &= \gamma_d(1+w)D \\ &= \left(113.2\ \frac{\text{lbf}}{\text{ft}^3}\right)(1+0.11)(8\ \text{ft}) \\ &= 1005\ \text{lbf/ft}^2\end{aligned}$$

Determine the shear strength at the midpoint of the clayey silty sand layer.

$$\begin{aligned}S &= c + \sigma\tan\phi \\ &= 250\ \frac{\text{lbf}}{\text{ft}^2} + \left(1005\ \frac{\text{lbf}}{\text{ft}^2}\right)\tan 31.7^\circ \\ &= 871\ \text{lbf/ft}^2 \quad (870\ \text{lbf/ft}^2)\end{aligned}$$

The answer is (D).

Why Other Options Are Wrong

(A) This incorrect solution can be obtained by multiplying the cohesion and the term for angle of internal friction.

(B) This incorrect answer results from using only the cohesion and not taking into consideration the angle of internal friction.

(C) This solution fails to convert the dry density to a total density.

SOLUTION 18

The time required for 90% consolidation of the clay layer can be found from the time rate of consolidation equation.

$$t = \frac{T_v H_d^2}{C_v}$$

Determined by Terzaghi's theory of consolidation, the time factor, T_v, that corresponds to 90% consolidation is typically taken as 0.848. Based on the given subsurface soil profile, the clay layer is double-drained with a drainage distance of 2.5 ft. Using the test results for C_v, the time required for 90% consolidation can be calculated by solving the time rate of consolidation equation for time.

$$\begin{aligned}t &= \frac{T_v H_d^2}{C_v} = \frac{(0.848)(2.5\ \text{ft})^2}{\left(7.3\times10^{-7}\ \frac{\text{ft}^2}{\text{sec}}\right)\left(86{,}400\ \frac{\text{sec}}{\text{day}}\right)} \\ &= 84\ \text{days} \quad (85\ \text{days})\end{aligned}$$

The answer is (C).

Why Other Options Are Wrong

(A) This incorrect solution uses a time factor of 0.197 (50% consolidation)—a mistake caused by misunderstanding how to obtain the correct time factor.

(B) This incorrect solution is obtained by assuming the clay layer is single-drained. Therefore, the drainage distance is doubled in the calculation. The mistake made here is misunderstanding how to apply the correct drainage distance.

(D) This incorrect solution is obtained using a single-drained layer—a mistake caused by misinterpreting the difference between a single- and a double-drained layer.

SOLUTION 19

The time required for 75% consolidation of the clay layer can be found from the time rate of consolidation equation.

$$T_v = \frac{tC_v}{H_d^2}$$

The time factor, T_v, required for 75% consolidation is typically taken as 0.477. Based on the subsurface soil profile, the clay layer is double-drained with a 3 ft long drainage path. Using C_v, the time required for 75% consolidation can be calculated.

$$\begin{aligned} t &= \frac{T_v H_d^2}{C_v} \\ &= \frac{(0.477)(3.0 \text{ ft})^2}{\left(1.3 \times 10^{-6} \ \frac{\text{ft}^2}{\text{sec}}\right)\left(86{,}400 \ \frac{\text{sec}}{\text{day}}\right)} \\ &= 38 \text{ days} \quad (40 \text{ days}) \end{aligned}$$

The answer is (B).

Why Other Options Are Wrong

(A) Failure to square the drainage path value in the calculation results in this answer, in which the units do not work out.

(C) This solution is incorrect and is caused by using a time factor of 0.848, the time factor associated with 90% consolidation.

(D) This incorrect solution is obtained by assuming a single-drained layer.

SOLUTION 20

The time required for 80% consolidation of the clay layer can be found from the time rate of consolidation equation.

$$T_v = \frac{tC_v}{H_d^2}$$

The time factor, T_v, required for 80% consolidation is typically taken as 0.567. Based on the subsurface soil profile, the clay layer is single-drained with a 30 ft long drainage path. Using C_v, the time required for 80% consolidation can be calculated.

$$\begin{aligned} t &= \frac{T_v H_d^2}{C_v} \\ &= \frac{(0.567)(30 \text{ ft})^2}{\left(4.3 \times 10^{-6} \ \frac{\text{ft}^2}{\text{sec}}\right)\left(86{,}400 \ \frac{\text{sec}}{\text{day}}\right)} \\ &= 1374 \text{ days} \quad (1400 \text{ days}) \end{aligned}$$

The answer is (C).

Why Other Options Are Wrong

(A) This solution fails to square the drainage path value in the calculation. The units do not work out.

(B) This incorrect solution assumes a double-drained layer.

(D) This solution is incorrect because it uses a time factor of 0.848 for 90% consolidation.

SOLUTION 21

The unit weight of the test sand can be used to determine the volume of the test hole in the field. The weight of dry soil obtained from the test hole can be divided by the volume of the test hole to determine the in-place dry unit weight.

The quantity of test sand needed by weight to fill the hole is given as 2.23 lbf. Determine the volume of the test hole using phase relationships.

$$\begin{aligned} V_{\text{test hole}} &= \frac{W_{\text{test sand}}}{\gamma_{\text{test sand}}} \\ &= \frac{2.23 \text{ lbf}}{84.7 \ \frac{\text{lbf}}{\text{ft}^3}} \\ &= 0.0263 \text{ ft}^3 \end{aligned}$$

The weight of dry soil obtained from the test hole can be calculated using the relationship

$$\begin{aligned} W_{\text{fill soil,dry}} &= \frac{W_{\text{fill soil,moist}}}{1 + w} \\ &= \frac{3.65 \text{ lbf}}{1 + 0.173} \\ &= 3.11 \text{ lbf} \end{aligned}$$

The in-place dry unit weight can be calculated by dividing the weight of dry soil obtained from the test hole by the volume of the test hole.

$$\begin{aligned}\gamma_{\text{fill soil,dry}} &= \frac{W_{\text{fill soil,dry}}}{V_{\text{test hole}}} \\ &= \frac{3.11\ \text{lbf}}{0.0263\ \text{ft}^3} \\ &= 118.3\ \text{lbf/ft}^3\end{aligned}$$

The in-place percent compaction is

$$\begin{aligned}\frac{\gamma_{\text{fill soil,dry}}}{\gamma_{\text{dry,max}}} \times 100\% &= \frac{118.3\ \frac{\text{lbf}}{\text{ft}^3}}{122.3\ \frac{\text{lbf}}{\text{ft}^3}} \times 100\% \\ &= 96.7\% \quad (97\%)\end{aligned}$$

The answer is (A).

Why Other Options Are Wrong

(B) This incorrect solution is obtained by dividing the maximum Proctor dry unit weight by the dry unit weight of the fill soil and multiplying by 100. The result is the reciprocal of the correct answer.

(C) This incorrect solution fails to convert the total weight of the fill soil to a dry weight. The incorrect value is then divided by the test hole volume to obtain a greater than correct unit weight and subsequent incorrect answer for percent compaction.

(D) This incorrect solution is obtained when the total weight of the fill soil is incorrectly converted to a dry weight (multiplying by $1 + w$ rather than dividing by it). The result is much greater than the correct answer.

3 Field Materials Testing, Methods, and Safety

BORROW SOURCE STUDIES

PROBLEM 1

A borrow source material, clayey silty sand, is proposed for construction of a 180,000 yd^3 embankment fill. Laboratory test results for the borrow source material are given.

moisture content (%)	moist unit weight (lbf/ft^3)	specific gravity	liquid limit	plasticity index
9.3	97.4	2.65	20	5

Specifications require the embankment to be compacted to an in-place dry density of 113.4 lbf/ft^3 and $\pm 2\%$ optimum moisture content. Approximate the quantity of borrow required to construct the embankment.

(A) 144,000 yd^3

(B) 198,000 yd^3

(C) 216,000 yd^3

(D) 230,000 yd^3

Hint: Use phase relationships to determine the weight of solids required to construct the embankment.

EXCAVATION AND EMBANKMENT

PROBLEM 2

A 14 ft deep trench is to be excavated in non-layered soil. From field observations, borehole collapse occurred at 5 ft below the ground surface. The soil was difficult to retain in a split-spoon sampler. From laboratory testing, the plasticity index (PI) of the fines in the soil sample was less than 4, 95% of the soil sample passed a no. 4 sieve, and 12% of the soil sample passed a no. 200 sieve.

According to OSHA, what is the maximum allowable slope for the trench walls?

(A) $\frac{3}{4}$:1

(B) 1:1

(C) $1\frac{1}{2}$:1

(D) 2:1

Hint: OSHA 29 CFR 1926 Subpart P gives requirements for excavation.

PROBLEM 3

A hospital with below-grade facilities is under construction in a location where a braced excavation to a depth of 37 ft will be required. The soils report indicates a clay profile with the parameters shown. Friction resistance can be ignored between the soils and between any future vertical retaining structures.

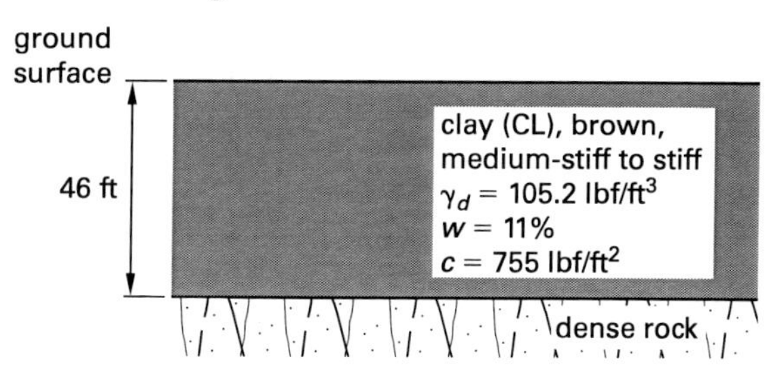

What is most nearly the expected maximum lateral pressure at a depth approximately at the midpoint of the excavation depth?

(A) 860 lbf/ft^2

(B) 1300 lbf/ft^2

(C) 2100 lbf/ft^2

(D) 3000 lbf/ft^2

Hint: Since an angle of internal friction was not given for the soil, assume the value is zero.

PROBLEM 4

A 6 ft wide by 20 ft deep trench is being excavated for utility installation. The soil profile and strength parameters are shown.

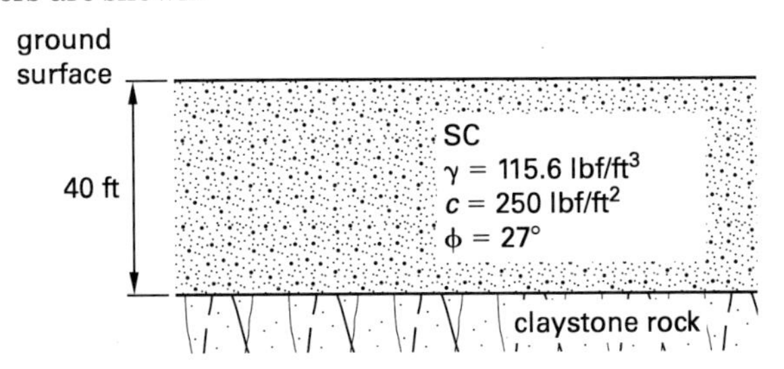

If struts are spaced every 30 ft on center, what is most nearly the expected maximum lateral force required?

(A) 70 kips

(B) 100 kips

(C) 170 kips

(D) 340 kips

Hint: The strut spacing results in a contributory width of 30 ft.

PROBLEM 5

A below-grade parking garage requires a braced excavation to a depth of 48 ft in the soil profile shown. A strut is proposed for every 12 ft of vertical depth.

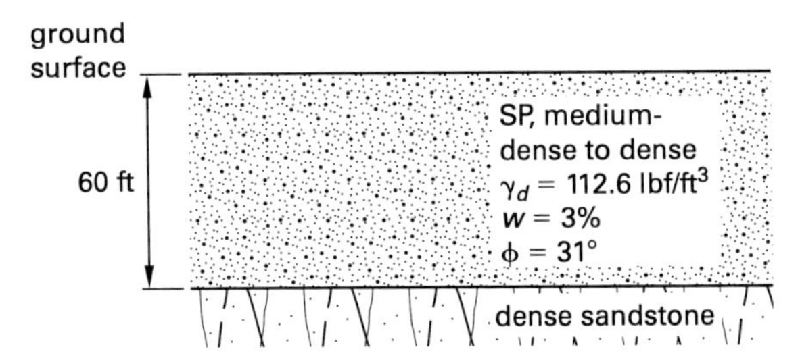

Use the Tschebotarioff trapezoidal earth pressure distribution to estimate the expected maximum lateral load on a strut, per unit width of bracing, at the midpoint of the excavation depth.

(A) 13.9 kips/ft

(B) 16.6 kips/ft

(C) 17.1 kips/ft

(D) 34.2 kips/ft

Hint: The struts are installed as the excavation proceeds downward.

PROBLEM 6

In the soil profile shown, a 25 ft deep excavation is to be cut for construction of a long pipeline.

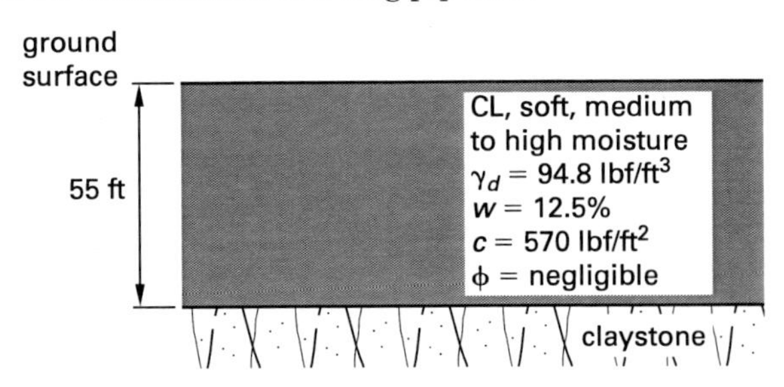

Approximate the allowable width of the cut required to maintain a minimum factor of safety of 1.5 against heave.

(A) 13 ft

(B) 25 ft

(C) 37 ft

(D) 50 ft

Hint: Because the trench is expected to be very long, the value for N_c can be used to estimate the maximum allowable width.

PROBLEM 7

A deep excavation is planned to the water table depth in the soil profile shown.

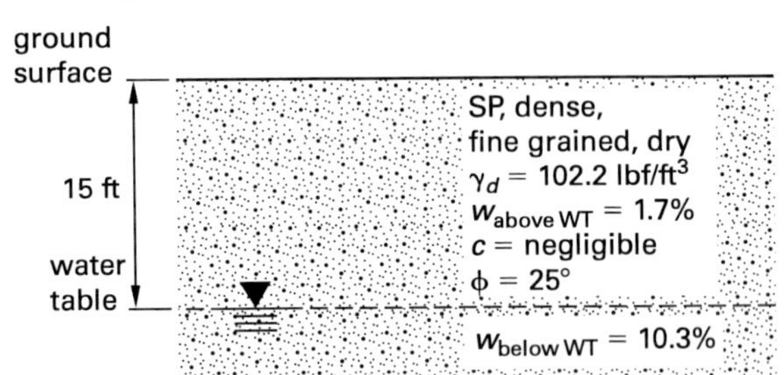

Given that the water table is stationary, approximate the factor of safety using Terzaghi bearing capacity factors.

(A) 1.8

(B) 2.8

(C) 3.4

(D) 3.7

Hint: Because the base of the cut is at the water table elevation and the water table is stationary, the density behind the cut is different than the density below the cut.

PROBLEM 8

A braced excavation is required for construction of a below-grade mat foundation in the soil profile shown.

The mat is to be 60 ft by 60 ft square, and the bottom of the excavation will require an additional 10 ft on both sides for purposes of constructing the foundation formwork.

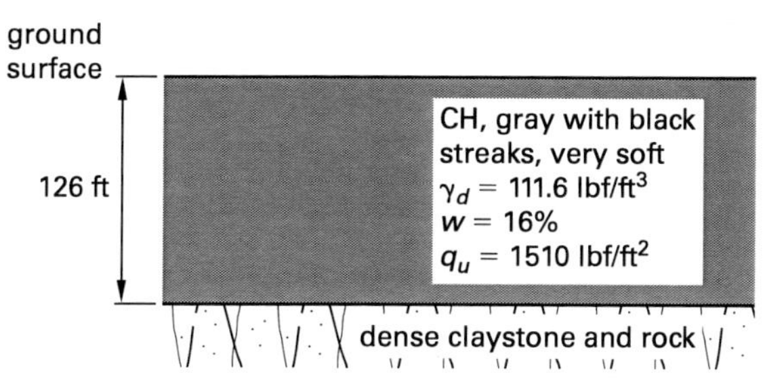

Approximate the maximum depth to which the cut can be extended without causing bottom heave.

(A) 33 ft

(B) 37 ft

(C) 44 ft

(D) 84 ft

Hint: The depth of the excavation can be expected to be less than the width.

LABORATORY AND FIELD COMPACTION

PROBLEM 9

A field technician obtains a soil sample from a clay liner under construction at a landfill closure project. Specifications require the clay liner to be compacted to a maximum dry density of 108.9 lbf/ft^3 and an optimum moisture content of 16%. If the initial moist sample weight is 5.0 lbf and the final dry sample weight after oven drying is 4.4 lbf, what is the difference between the field moisture content and the specified optimum moisture content of the liner material?

(A) 2.0%

(B) 4.0%

(C) 14%

(D) 72%

Hint: Find the water content of the clay liner soil in the field.

PROBLEM 10

The dry density of a saturated silty sand in the field is 111.1 lbf/ft^3, the specific gravity of the solids is 2.65, and the plastic limit (PL) is 13. The difference on a percent solids basis between the water content of the saturated field sample and the PL of the soil is most nearly

(A) 5.5%

(B) 18%

(C) 25%

(D) 38%

Hint: The water content of the soil at the plastic limit is the PL.

Field Testing, Methods, Safety

SOLUTION 1

The weight of solids per unit total volume required for the embankment is the same as the weight of solids to be removed per unit volume from the borrow source. Therefore, the weight of solids required for the embankment may be used to determine the quantity, in cubic yards, of the borrow source soil required to construct the embankment. Calculate the weight of solids based on the total embankment volume.

$$\begin{aligned} V_{t,\text{emb}} &= (180{,}000\ \text{yd}^3)\left(3\ \frac{\text{ft}}{\text{yd}}\right)^3 \\ &= 4.86 \times 10^6\ \text{ft}^3 \end{aligned}$$

$$\gamma_{d,\text{emb}} = \frac{W_{s,\text{emb}}}{V_{t,\text{emb}}}$$

$$\begin{aligned} W_{s,\text{emb}} &= \gamma_{d,\text{emb}} V_{t,\text{emb}} \\ &= \left(113.4\ \frac{\text{lbf}}{\text{ft}^3}\right)(4.86 \times 10^6\ \text{ft}^3) \\ &= 5.51 \times 10^8\ \text{lbf} \end{aligned}$$

The total weight of soil required from the borrow source can be determined from the weight of soil required for the embankment.

$$\begin{aligned} W_{\text{total,borrow}} &= W_{s,\text{emb}}(1 + w_{\text{borrow}}) \\ &= (5.51 \times 10^8\ \text{lbf})(1 + 0.093) \\ &= 6.02 \times 10^8\ \text{lbf} \end{aligned}$$

The volume required from the borrow source can be determined.

$$\begin{aligned} V_{\text{total,borrow}} &= \left(\frac{W_{\text{total,borrow}}}{\gamma_{t,\text{borrow}}}\right)\left(\frac{1\ \text{yd}}{3\ \text{ft}}\right)^3 \\ &= \left(\frac{6.02 \times 10^8\ \text{lbf}}{97.4\ \frac{\text{lbf}}{\text{ft}^3}}\right)\left(\frac{1\ \text{yd}}{3\ \text{ft}}\right)^3 \\ &= 229{,}060\ \text{yd}^3\ \text{of borrow} \quad (230{,}000\ \text{yd}^3) \end{aligned}$$

The answer is (D).

Why Other Options Are Wrong

(A) This solution mistakenly calculates the cubic yardage of fill required per cubic yard of borrow material and then finds the wrong quantity of borrow material.

(B) Mistakenly multiplying the moist unit weight and the moisture content of the borrow source rather than dividing results in this incorrect answer.

(C) If the moist unit weight is used instead of the dry unit weight to determine the required quantity of borrow soils, this incorrect result is obtained.

SOLUTION 2

OSHA 29 CFR 1926.652 and its Apps. A and B cover the maximum allowable slope based on soil type for trench walls.

The soil classification system defined in OSHA 29 CFR 1926 Subpart P, App. A uses soil texture and grain size to classify soil into three types: A, B, or C. Since less than 50% of the soil passed the no. 200 sieve, and more than 50% of the coarse fraction passed the no. 4 sieve, the soil is coarse-grained and primarily contains sand.

Type A soil includes relatively stable, cohesive and plastic soils that are primarily composed of clay. However, the sample soil is unstable because the boreholes collapsed and because a sample was difficult to retain with a split-spoon sampler. The soil cannot be type A.

Type B soil includes certain granular, cohesionless soils with no plasticity, such as soils that are primarily composed of silt or loam. The fines in the sample have little to no plasticity, which can indicate that the fines are silty or loamy. However, the sample soil is coarse-grained, so it cannot be type B.

Type C soil includes soils that are primarily composed of sand or gravel. Since the sample is primarily coarse-grained, it is type C soil. While the lab result indicating a plasticity index (PI) of less than 4 confirms that the sample contains low fines, this data is not necessary to determine the OSHA soil type.

The planned trench depth is 14 ft. OSHA table B-1 specifies that for type C soil and for trenches between 12 ft and 20 ft, the maximum allowable slope is 1½:1.

The answer is (C).

Why Other Options Are Wrong

(A) This incorrect option is the allowable slope for type A soils. Type C soil cannot support this slope.

(B) This incorrect option is the allowable slope for type B soils. Type C soil cannot support this slope.

(D) This incorrect option does not specify the maximum allowable slope that type C soil can support.

SOLUTION 3

The maximum horizontal earth pressure at the midpoint of the bracing can be calculated.

$$\begin{aligned} p_{\max} &= k_a \gamma H \\ &= k_a \gamma_d (1 + w) H \end{aligned}$$

The Rankine theory should not be used to find the active earth pressure coefficient for the soil because the excavation is to be braced as the construction progresses downward into the soil mass. The horizontal earth pressure coefficient can be calculated.

$$\begin{aligned} k_a &= 1 - \frac{4c}{\gamma H} \\ &= 1 - \frac{4c}{\gamma_d(1+w)H} \\ &= 1 - \frac{(4)\left(755\ \frac{\text{lbf}}{\text{ft}^2}\right)}{\left(105.2\ \frac{\text{lbf}}{\text{ft}^3}\right)(1+0.11)(37\ \text{ft})} \\ &= 0.30 \end{aligned}$$

The maximum horizontal earth pressure at the midpoint of the bracing can be calculated.

$$\begin{aligned} p_{\text{max}} &= k_a\gamma_d(1+w)H \\ &= (0.30)\left(105.2\ \frac{\text{lbf}}{\text{ft}^3}\right)(1+0.11)(37\ \text{ft}) \\ &= 1296\ \text{lbf/ft}^2 \quad (1300\ \text{lbf/ft}^2) \end{aligned}$$

The answer is (B).

Why Other Options Are Wrong

(A) This solution mistakenly omits the adjustment for moisture content in both calculations for the horizontal earth pressure and coefficient.

(C) This solution mistakenly uses the Rankine theory for determination of the horizontal active earth pressure at the base of the excavation.

(D) This solution fails to subtract the second term from one in the calculation for the horizontal earth pressure coefficient.

SOLUTION 4

The maximum horizontal earth pressure at the midpoint of the bracing should be calculated using the typical rectangular-shaped earth pressure distribution used for braced cuts in drained sand (SC, clayey sand).

$$p_{\text{max}} = 0.65k_a\gamma H$$

Calculate the active earth pressure coefficient for the soil.

$$\begin{aligned} k_a &= \tan^2\left(45° - \frac{\phi}{2}\right) \\ &= \tan^2\left(45° - \frac{27°}{2}\right) \\ &= 0.38 \end{aligned}$$

The maximum horizontal earth pressure at the midpoint of the bracing can be calculated.

$$\begin{aligned} p_{\text{max}} &= 0.65k_a\gamma H \\ &= (0.65)(0.38)\left(115.6\ \frac{\text{lbf}}{\text{ft}^3}\right)(20\ \text{ft}) \\ &= 571\ \text{lbf/ft}^2 \end{aligned}$$

Based on the strut spacing of 30 ft and the trench depth of 20 ft, the resulting maximum horizontal force can be calculated.

$$\begin{aligned} R_{\text{max}} &= D(\text{strut spacing})p_{\text{max}} \\ &= (20\ \text{ft})(30\ \text{ft})\left(571\ \frac{\text{lbf}}{\text{ft}^2}\right) \\ &= 342{,}600\ \text{lbf} \quad (340\ \text{kips}) \end{aligned}$$

The answer is (D).

Why Other Options Are Wrong

(A) This solution mistakenly uses the trench width instead of the contributory width.

(B) This solution mistakenly uses the trench width instead of the depth.

(C) This solution mistakenly uses half the contributory width.

SOLUTION 5

A strut will be placed at the midpoint depth of 24 ft in the excavation. The Tschebotarioff trapezoidal pressure distribution can be used to calculate the maximum lateral pressure at $0.5H$.

$$p_{\text{max}} = 0.8k_a\gamma H$$

Determine the active earth pressure coefficient for the soil (SP, pooly graded sand).

$$\begin{aligned} k_a &= \frac{1-\sin\phi}{1+\sin\phi} = \frac{1-\sin 31°}{1+\sin 31°} \\ &= 0.32 \end{aligned}$$

The maximum horizontal earth pressure at the midpoint of the bracing can be calculated.

$$\begin{aligned} p_{\max} &= 0.8k_a\gamma_d(1+w)H \\ &= (0.8)(0.32)\left(112.6\ \frac{\text{lbf}}{\text{ft}^3}\right)(1+0.03)(48\ \text{ft}) \\ &= 1425\ \text{lbf/ft}^2 \end{aligned}$$

The strut will support a contributory area consisting of a contributory depth per foot of bracing support width. Calculate the load on the strut per foot of bracing width.

$$\begin{aligned} R_{\text{max at 24 ft}} &= p_{\max}A_{\text{contributory}} \\ &= p_{\max}D_{\text{contributory}}w_{\text{unit bracing}} \\ &= \left(1425\ \frac{\text{lbf}}{\text{ft}^2}\right)(12\ \text{ft}) \\ &= 17{,}100\ \text{lbf/ft of bracing width} \\ &\quad (17.1\ \text{kips/ft}) \end{aligned}$$

The answer is (C).

Why Other Options Are Wrong

(A) This solution mistakenly uses the "typical" earth pressure distribution instead of what is specifically requested, Tschebotarioff, to calculate the maximum pressure.

(B) This solution mistakenly omits the adjustment for moisture content in the calculation for the horizontal earth pressure.

(D) This solution mistakenly calculates the contributory area consisting of a contributory depth of 24 ft instead of 12 ft per foot of bracing support width.

SOLUTION 6

Based on the given strength parameters, soil type (CL, clay), and indication of soft in-situ conditions, assume the depth of the excavation will be larger than the width. Use the equation for the factor of safety to solve for N_c.

$$\begin{aligned} F &= \frac{N_c c}{\gamma H + q} \\ N_c &= \frac{F(\gamma H + q)}{c} \\ &= \frac{F(\gamma_d(1+w)H + q)}{c} \\ &= \frac{(1.5)\left(\left(94.8\ \frac{\text{lbf}}{\text{ft}^3}\right)(1+0.125)(25\ \text{ft}) + 0\right)}{570\ \frac{\text{lbf}}{\text{ft}^2}} \\ &= 7.0 \end{aligned}$$

Confirm that the excavation depth is less than the critical height.

$$\begin{aligned} H_c &= \frac{N_c c}{\gamma} = \frac{N_c c}{\gamma_d(1+w)} \\ &= \frac{(7.0)\left(570\ \frac{\text{lbf}}{\text{ft}^2}\right)}{\left(94.8\ \frac{\text{lbf}}{\text{ft}^3}\right)(1+0.125)} \\ &= 37.4\ \text{ft} \quad (37\ \text{ft}) \end{aligned}$$

$$25\ \text{ft} < 37\ \text{ft}$$

The excavation depth is less than the critical height. Use the value for N_c to determine the allowable excavation width in reference to published correlations. For N_c equal to 7, the maximum allowable width of the excavation can be determined.

$$\begin{aligned} \frac{H}{B} &= 2 \\ B &= \frac{H}{2} = \frac{25\ \text{ft}}{2} \\ &= 12.5\ \text{ft} \quad (13\ \text{ft}) \end{aligned}$$

The answer is (A).

Why Other Options Are Wrong

(B) This solution miscalculates the value for N_c by failing to convert the dry unit weight to a moist unit weight. The result is used to determine the allowable excavation width.

(C) This solution mistakenly calculates the critical height instead of the allowable width.

(D) This solution miscalculates the allowable width by incorrectly inverting the value of H/B that corresponds to the value for N_c.

SOLUTION 7

The factor of safety for the stability of the excavation can be calculated using an equation for the depth equivalent to the stationary water table depth.

$$F = 2N_\gamma\left(\frac{\gamma_{\text{sub}}}{\gamma_{\text{drained}}}\right)k_a \tan\phi$$

The value for the bearing capacity factor, $N_\gamma = 9.7$, can be obtained from published sources. Calculate the coefficient of active horizontal earth pressure.

$$\begin{aligned} k_a &= \frac{1-\sin\phi}{1+\sin\phi} = \frac{1-\sin 25^\circ}{1+\sin 25^\circ} \\ &= 0.41 \end{aligned}$$

Calculate the factor of safety.

$$\begin{aligned} F &= 2N_\gamma\left(\frac{\gamma_{\text{sub}}}{\gamma_{\text{drained}}}\right)k_a \tan\phi \\ &= 2N_\gamma\left(\frac{\gamma_d(1+w_{\text{below WT}})-\gamma_w}{\gamma_d(1+w_{\text{above WT}})}\right)k_a \tan\phi \\ &= (2)(9.7)\left(\frac{\left(102.2\ \frac{\text{lbf}}{\text{ft}^3}\right)(1+0.103) - 62.4\ \frac{\text{lbf}}{\text{ft}^3}}{\left(102.2\ \frac{\text{lbf}}{\text{ft}^3}\right)(1+0.017)}\right) \\ &\quad \times (0.41)\tan 25^\circ \\ &= 1.8 \end{aligned}$$

The answer is (A).

Why Other Options Are Wrong

(B) This solution neglects to use the Terzaghi bearing capacity factor, $N_\gamma = 9.7$, and instead uses the Meyerhof factor, $N_\gamma = 6.8$.

(C) This solution erroneously inverts the unit weight values in the numerator and the denominator.

(D) This solution neglects to use the appropriate equation for factor of safety. Instead, it uses the equation for an excavation that is conducted without groundwater influence.

SOLUTION 8

The critical height can be calculated for an excavation where the depth is expected to be less than the width.

The cohesion can be assumed to be half the value of the unconfined compressive strength.

$$\begin{aligned} c &= \frac{q_u}{2} = \frac{1510\ \frac{\text{lbf}}{\text{ft}^2}}{2} \\ &= 755\ \text{lbf/ft}^2 \end{aligned}$$

$$\begin{aligned} H_c &= \frac{5.7c}{\gamma - \sqrt{2}\left(\frac{c}{B}\right)} \\ &= \frac{5.7c}{\gamma_d(1+w) - \sqrt{2}\left(\frac{c}{B}\right)} \\ &= \frac{(5.7)\left(755\ \frac{\text{lbf}}{\text{ft}^2}\right)}{\left(111.6\ \frac{\text{lbf}}{\text{ft}^3}\right)(1+0.16) - \sqrt{2}\left(\frac{755\ \frac{\text{lbf}}{\text{ft}^2}}{80\ \text{ft}}\right)} \\ &= 37\ \text{ft} \end{aligned}$$

The answer is (B).

Why Other Options Are Wrong

(A) This solution mistakenly inverts the cohesion and excavation width in the equation.

(C) This solution omits the adjustment for moisture content in the calculation.

(D) This solution neglects to convert the unconfined compressive strength value to cohesion.

SOLUTION 9

Find the difference between the field measured water content and the specified optimum moisture content of the clay liner soil.

$$\begin{aligned} w_{\text{field}} &= \frac{W_m - W_d}{W_d} = \frac{5.0\ \text{lbf} - 4.4\ \text{lbf}}{4.4\ \text{lbf}} \times 100\% \\ &= 14\% \\ w_{\text{difference}} &= \text{OM} - w_{\text{field}} = 16\% - 14\% \\ &= 2.0\% \end{aligned}$$

The answer is (A).

Why Other Options Are Wrong

(B) This solution mistakenly uses the initial moist sample weight in the denominator when calculating the water content, rather than using the weight of the soil after oven drying. It then uses that incorrect water content value to find the difference between the optimum moisture content and the field water content.

(C) This incorrect solution is the field moisture content itself.

(D) This solution mistakenly uses the initial moist sample weight in the numerator when calculating the water content, rather than using the difference in initial soil weight and soil weight after oven drying. It then uses that incorrect water content value to find the difference between the optimum moisture content and the field water content.

SOLUTION 10

Find the difference between the water content of the field sample and the given PL of the soil. The field sample is saturated as the voids are completely filled by water. Therefore, the degree of saturation can be expressed using the following relationship.

$$S = \frac{w(\text{SG})}{e}$$

The expression for degree of saturation can be solved for the water content.

$$w = \frac{Se}{\text{SG}}$$

The dry density is given in the problem statement. The dry density can be expressed in terms of the specific gravity, void ratio, and unit weight of water.

$$\gamma_d = \frac{(\text{SG})\gamma_w}{1+e}$$

The expression for dry density can be solved for the void ratio.

$$e = \frac{(\text{SG})\gamma_w}{\gamma_d} - 1$$

The void ratio can be calculated.

$$e = \frac{(\text{SG})\gamma_w}{\gamma_d} - 1 = \frac{(2.65)\left(62.4\ \frac{\text{lbf}}{\text{ft}^3}\right)}{111.1\ \frac{\text{lbf}}{\text{ft}^3}} - 1$$
$$= 0.49$$

Using the value of 1 for the degree of saturation (100%), the water content can be calculated.

$$w = \frac{Se}{\text{SG}} = \frac{(1)(0.49)}{2.65}$$
$$= 0.185 \quad (18.5\%)$$

The water content at the plastic limit is the PL expressed as a percentage.

$$\text{PL} = 13 = 13\%$$

Calculate the difference between water content of the field sample and PL of the field sample.

$$\begin{aligned} w_{\text{diff}} &= w - w_{\text{PL}} \\ &= 18.5\% - 13\% \\ &= 5.5\% \end{aligned}$$

The answer is (A).

Why Other Options Are Wrong

(B) This solution mistakenly calculates the water content of the saturated sample without subtracting the PL.

(C) This solution mistakenly uses the volume of the solids instead of the volume of the voids as equivalent to the volume of water when determining the water content. It then subtracts the PL water content value. This answer is greater than the correct answer.

(D) This solution mistakenly uses the volume of the solids instead of the volume of the voids as equivalent to the volume of water when determining the water content. It also neglects to subtract the PL value.

4 Earthquake Engineering and Dynamic Loads

LIQUEFACTION

PROBLEM 1

Based on the soil profile shown and additional bearing pressure from a proposed 5 ft thick fill of large area placed at the ground surface, approximate the cyclic stress ratio (CSR) at the midpoint of the submerged sand layer. Assume a maximum peak acceleration of $0.45g$ is experienced at the ground surface above the sand layer due to a seismic event. The stress reduction factor has been calculated to be 0.93.

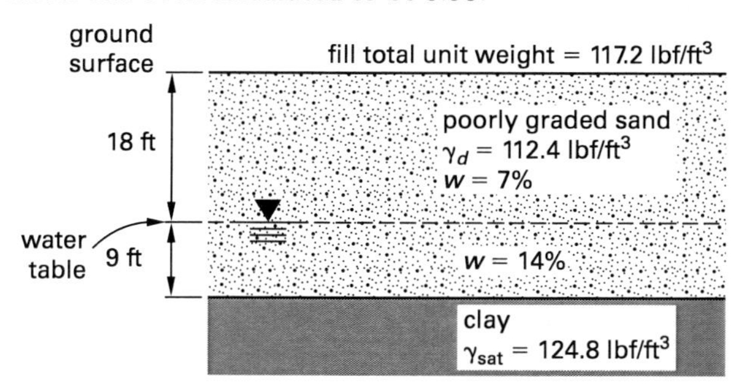

(A) 0.25

(B) 0.30

(C) 0.32

(D) 1.3

Hint: The CSR is based on the average effective stress of the liquefiable soil layer.

PROBLEM 2

Commencing at a depth of 11.5 ft, the uncorrected standard penetration test (SPT) results per 6 in interval in the liquefiable soil layer shown are 9, 10, and 8, respectively.

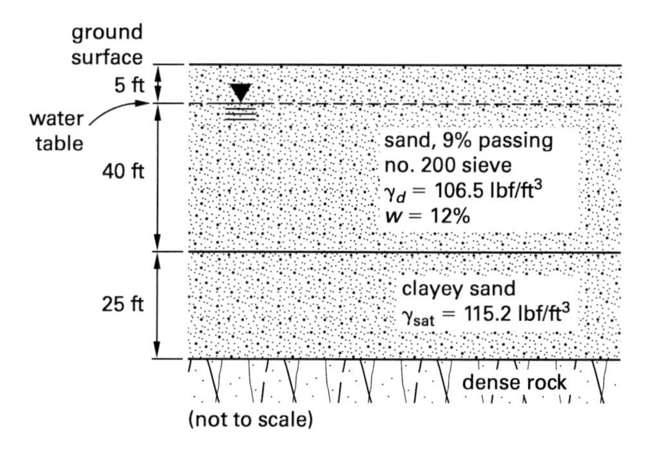

Approximate the cyclic resistance ratio inducible by a moment magnitude 7.5 earthquake ($CRR_{M=7.5}$) for a depth of 12 ft below the ground surface.

(A) 0.22

(B) 0.24

(C) 0.29

(D) 0.31

Hint: The standard penetration test data needs to be corrected for effective overburden pressure and fines content (normalized for "clean sand").

PROBLEM 3

For the liquefiable soil layers shown, determine the cyclic resistance ratio inducible by a moment magnitude 7.5 earthquake ($CRR_{M=7.5}$) for a depth of 16 ft below the ground surface. Assume atmospheric pressure is 14.7 lbf/in^2.

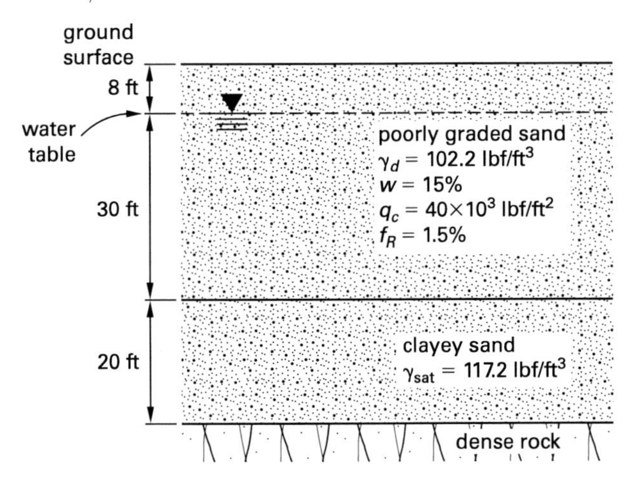

(A) 0.07

(B) 0.08

(C) 0.09

(D) 0.1

Hint: The cone penetration test (CPT) data needs to be normalized to 1 atm of pressure.

PROBLEM 4

Based on the soil profile shown and the following volumetric strain graph, estimate the vertical settlement after an earthquake of magnitude 7.5. Assume a maximum peak acceleration of $0.35g$ is experienced at the ground surface above the sand layer due to the seismic event. The stress reduction factor has been calculated to be 0.95.

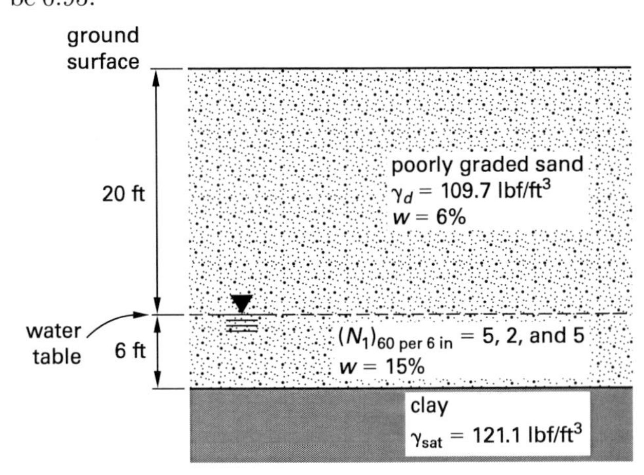

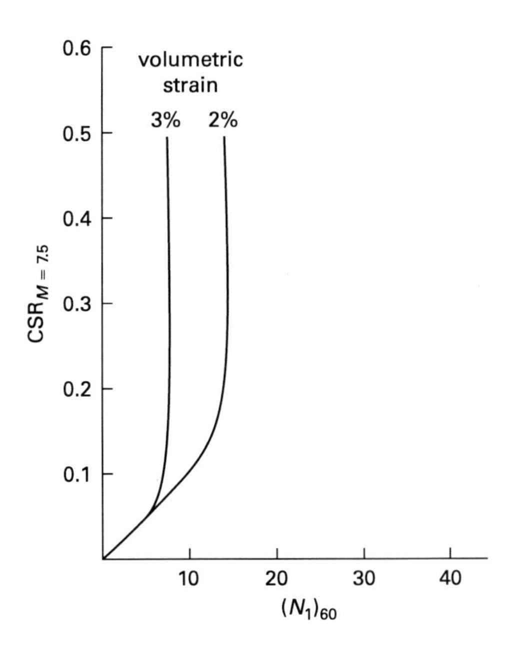

(A) 0.18 in

(B) 1.4 in

(C) 2.2 in

(D) 3.0 in

Hint: The vertical settlement can be estimated using the cyclic stress ratio for a magnitude 7.5 earthquake ($CSR_{M=7.5}$).

PSEUDO-STATIC ANALYSIS AND EARTHQUAKE LOADS

PROBLEM 5

A telecommunications tower under construction near a great fault has a total weight of 35 kips. If the spectral acceleration expected for the area is 0.43 ft/sec^2, determine the expected theoretical base shear.

(A) 0.47 lbf

(B) 470 lbf

(C) 2100 lbf

(D) 15,000 lbf

Hint: The base shear is based on a percentage of gravity.

PROBLEM 6

A sign weighing 16 kips is supported by three metal posts with the properties shown.

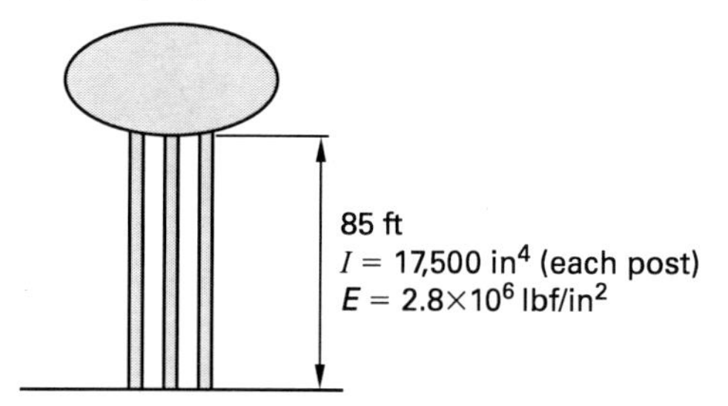

If the weights of the metal posts are negligible, approximate the natural frequency of the structure.

(A) 0.30 Hz

(B) 0.50 Hz

(C) 2.0 Hz

(D) 3.4 Hz

Hint: The stiffnesses of multiple posts can be summed.

SEISMIC SITE CHARACTERIZATION

PROBLEM 7

A seismometer station is located approximately 100 km from a known active fault. During an earthquake, the seismograph records the amplitudes shown.

time index (sec)	amplitude (mm)
10	5
12	5
14	7
16	10
18	14
20	20
22	22
24	25
26	23
28	15
30	10
32	5

Approximate the Richter magnitude of the earthquake.

(A) 3.7

(B) 4.0

(C) 4.2

(D) 4.4

Hint: The equation for determining Richter magnitude is applicable if the distance from the epicenter to the seismometer station is approximately 100 km.

PROBLEM 8

Calculate the approximate length of an active fault after an earthquake of Richter magnitude 3.8 is measured. Use an approximate expression for correlation to historic data.

(A) 0.01 m

(B) 0.1 m

(C) 10 m

(D) 100 m

Hint: An equation that approximates the length of an active fault based on Richter magnitude is available based on correlation with historic data.

PROBLEM 9

A standard penetration test (SPT) to a depth of 100 ft is conducted at a construction site. The tabulated test results are shown.

soil layer (top to bottom)	soil layer thickness, d (ft)	standard penetration resistance, N (blows/ft)
1	15	12
2	20	16
3	30	15
4	10	19
5	25	17

According to ASCE/SEI7 guidelines, what is the most appropriate site class?

(A) class C, very dense soil and soft rock

(B) class D, stiff soil

(C) class E, soft clay soil

(D) class F, soil requires a site response analysis

Hint: Refer to ASCE/SEI7 Sec. 20.4 to determine the site class.

SOLUTION 1

To determine the cyclic stress ratio (CSR), calculate the total overburden and effective stress.

$$\sigma_o = \sigma_o' + u$$
$$\sigma_o = \sigma_{\text{fill}} + \sigma_{\text{dry sand}} + \sigma_{\text{sub sand}}$$
$$\begin{aligned}\sigma_o &= \gamma_{\text{fill}} D_{\text{fill}} + \gamma D_{\text{dry sand}} + \gamma D_{\text{sub sand}} \\ &= \gamma_{\text{fill}} D_{\text{fill}} + \gamma_d (1 + w_{\text{dry sand}}) D_{\text{dry sand}} \\ &\quad + \gamma_d (1 + w_{\text{sub sand}}) D_{\text{sub sand}} \\ &= \left(117.2\ \frac{\text{lbf}}{\text{ft}^3}\right)(5\text{ ft}) + \left(112.4\ \frac{\text{lbf}}{\text{ft}^3}\right) \\ &\quad \times (1+0.07)(18\text{ ft}) + \left(112.4\ \frac{\text{lbf}}{\text{ft}^3}\right) \\ &\quad \times (1+0.14)(4.5\text{ ft}) \\ &= 3327\text{ lbf/ft}^2\end{aligned}$$

$$\begin{aligned}\sigma_o' &= \sigma_o - \mu \\ &= 3327\ \frac{\text{lbf}}{\text{ft}^2} - \left(62.4\ \frac{\text{lbf}}{\text{ft}^3}\right)(4.5\text{ ft}) \\ &= 3046\text{ lbf/ft}^2\end{aligned}$$

Calculate the CSR.

$$\begin{aligned}\frac{\tau_{h,\text{ave}}}{\sigma_o'} &= \text{CSR} \\ &= 0.65\left(\frac{a_{\max}}{g}\right)\left(\frac{\sigma_o}{\sigma_o'}\right) r_d \\ &= (0.65)(0.45)\left(\frac{3327\ \frac{\text{lbf}}{\text{ft}^2}}{3046\ \frac{\text{lbf}}{\text{ft}^2}}\right)(0.93) \\ &= 0.30\end{aligned}$$

The answer is (B).

Why Other Options Are Wrong

(A) This solution mistakenly calculates the CSR by using the reciprocal of the effective stress and the total stress ratio.

(C) This solution mistakenly calculates the effective stress at the base of the liquefiable layer rather than at the midpoint (average).

(D) This solution mistakenly ignores the fill as an additional stress and calculates the CSR without adding in the stress due to the fill.

SOLUTION 2

The cyclic resistance ratio ($\text{CRR}_{M=7.5}$) can be calculated from standard penetration test (SPT) data.

$$\begin{aligned}\text{CRR}_{M=7.5} &= \frac{1}{34 - (N_1)_{60,\text{cs}}} + \frac{(N_1)_{60,\text{cs}}}{135} \\ &\quad + \frac{50}{\left(10(N_1)_{60,\text{cs}} + 45\right)^2} - \frac{1}{200}\end{aligned}$$

Normalize the SPT data for “clean sand.”

$$(N_1)_{60,\text{cs}} = \alpha + \beta (N_1)_{60}$$

For a fines content less than 35%,

$$\alpha = e^{1.76 - \frac{190}{\text{fines}^2}}$$
$$\beta = 0.99 + \frac{\text{fines}^{1.5}}{1000}$$

To calculate $\text{CRR}_{M=7.5}$, the SPT data for the depth of 12 ft should be corrected for effective overburden pressure and fines content. The correction uses a comparison of the initial effective stress to atmospheric pressure.

$$\begin{aligned}N_{60} &= N_{60,\text{2nd 6 in}} + N_{60,\text{3rd 6 in}} \\ &= 10 + 8 \\ &= 18\end{aligned}$$

$$(N_1)_{60} = N_{60}\left(\frac{2.2}{1.2 + \frac{\sigma_{vo}'}{P_{\text{atm}}}}\right)$$

$$\begin{aligned}\sigma_{vo}' &= \gamma_{d,\text{sand}} D_{\text{dry sand}} + \gamma_{d,\text{sand}} \\ &\quad \times (1 + w) D_{\text{wet sand at 12 ft}} - \gamma_w D_{w\text{ at 12 ft}} \\ &= \left(106.5\ \frac{\text{lbf}}{\text{ft}^3}\right)(5\text{ ft}) + \left(106.5\ \frac{\text{lbf}}{\text{ft}^3}\right) \\ &\quad \times (1+0.12)(7\text{ ft}) - \left(62.4\ \frac{\text{lbf}}{\text{ft}^3}\right)(7\text{ ft}) \\ &= 930.7\text{ lbf/ft}^2\end{aligned}$$

Earthquake Eng. Dynamic Loads

$$(N_1)_{60} = N_{60}\left(\frac{2.2}{1.2+\frac{\sigma_{vo}}{P_{atm}}}\right)$$

$$= (18)\left(\frac{2.2}{1.2+\frac{930.7\ \frac{\text{lbf}}{\text{ft}^2}}{2000\ \frac{\text{lbf}}{\text{ft}^2}}}\right)$$

$$= 24$$

$$(N_1)_{60,\text{cs}} = \alpha + \beta(N_1)_{60}$$

$$= e^{1.76-\frac{190}{\text{fines}^2}} + \left(0.99+\frac{\text{fines}^{1.5}}{1000}\right)(N_1)_{60}$$

$$= e^{1.76-\frac{190}{(9)^2}} + \left(0.99+\frac{(9)^{1.5}}{1000}\right)(24)$$

$$= 25$$

The $\text{CRR}_{M=7.5}$ can be calculated.

$$\text{CRR}_{M=7.5} = \frac{1}{34-(N_1)_{60,\text{cs}}} + \frac{(N_1)_{60,\text{cs}}}{135} + \frac{50}{(10(N_1)_{60,\text{cs}}+45)^2} - \frac{1}{200}$$

$$= \frac{1}{34-25} + \frac{25}{135} + \frac{50}{((10)(25)+45)^2} - \frac{1}{200}$$

$$= 0.29$$

The answer is (C).

Why Other Options Are Wrong

(A) This solution fails to correct the SPT data for overburden pressure and fines content.

(B) This solution fails to determine the effective stress and instead uses the total stress in calculating the correction of the SPT data for overburden pressure.

(D) This solution miscalculates the SPT data by adding the first two values rather than ignoring the first number and adding the last two values.

SOLUTION 3

The cyclic resistance ratio ($\text{CRR}_{M=7.5}$) can be calculated from the tip penetration resistance portion of the cone penetration test (CPT) data after normalization to an equivalent "clean sand" value and 1 atm of pressure.

Based on plotted curve data, the applicable equation should be determined.

$$\text{CRR}_{M=7.5} = (0.833)\left(\frac{(q_{c1\text{N}})}{1000}\right) + 0.05 \quad [\text{for } (q_{c1\text{N}})_{\text{CS}} < 50]$$

$$\text{CRR}_{M=7.5} = (93)\left(\frac{(q_{c1\text{N}})}{1000}\right)^3 + 0.08 \quad [\text{for } 50 \le (q_{c1\text{N}})_{\text{CS}} < 160]$$

The factor, C_Q, is for normalizing the tip penetration data to 1 atm of pressure. The normalized tip penetration data should be corrected for grain characteristics using the factor, K_c. The exponent, n, varies between 0.5 and 1.0 for different soil types.

$$(q_{c1\text{N}})_{\text{CS}} = K_c q_{c1\text{N}}$$

$$q_{c1\text{N}} = C_Q\left(\frac{q_c}{P_{\text{atm}}}\right)$$

$$C_Q = \left(\frac{P_{\text{atm}}}{\sigma'_{vo}}\right)^n \le 1.7$$

Based on Robertson and Wride soil behavior data and corresponding relationships, calculate the CPT soil behavior type index. An iteration procedure using the exponent, n, is used with the soil behavior type index to normalize the tip penetration data of sand with fines to that of a clean sand value.

$$I_c = \sqrt{(3.47-\log Q)^2 + (1.22+\log F)^2}$$

$$Q = \left(\frac{q_c-\sigma_{vo}}{P_{\text{atm}}}\right)\left(\frac{P_{\text{atm}}}{\sigma'_{vo}}\right)^n$$

$$F = \frac{f_s}{q_c-\sigma_{vo}} \times 100\%$$

Determine the total stress, effective stress, and CPT soil behavior type index by assuming an initial value of $n = 1$.

$$\sigma_{vo} = \gamma_{d,\text{sand}} D_{\text{dry sand}} + \gamma_{d,\text{sand}}(1+w) \times D_{\text{wet sand at 16 ft}}$$

$$= \left(102.2\ \frac{\text{lbf}}{\text{ft}^3}\right)(8\text{ ft}) + \left(102.2\ \frac{\text{lbf}}{\text{ft}^3}\right) \times (1+0.15)(8\text{ ft})$$

$$= 1758\ \text{lbf/ft}^2$$

$$\mu = \gamma_w D_w = (8\text{ ft})\left(62.4\ \frac{\text{lbf}}{\text{ft}^3}\right)$$

$$= 499\ \text{lbf/ft}^2$$

Earthquake Eng. Dynamic Loads

$$\sigma'_{vo} = \sigma_{vo} - \mu = 1758\ \frac{\text{lbf}}{\text{ft}^2} - 499\ \frac{\text{lbf}}{\text{ft}^2} = 1259\ \text{lbf/ft}^2$$

$$Q = \left(\frac{q_c - \sigma_{vo}}{p_{\text{atm}}}\right)\left(\frac{p_{\text{atm}}}{\sigma'_{vo}}\right)^n = \left(\frac{40{,}000\ \frac{\text{lbf}}{\text{ft}^2} - 1758\ \frac{\text{lbf}}{\text{ft}^2}}{2117\ \frac{\text{lbf}}{\text{ft}^2}}\right)\left(\frac{2117\ \frac{\text{lbf}}{\text{ft}^2}}{1259\ \frac{\text{lbf}}{\text{ft}^2}}\right)^1 = 30.4$$

$$F = \frac{f_s}{q_c - \sigma_{vo}} \times 100\% = \frac{f_R q_c}{q_c - \sigma_{vo}} \times 100\% = \frac{(0.015)\left(40{,}000\ \frac{\text{lbf}}{\text{ft}^2}\right)}{40{,}000\ \frac{\text{lbf}}{\text{ft}^2} - 1758\ \frac{\text{lbf}}{\text{ft}^2}} \times 100\% = 1.57$$

Calculate the CPT soil behavior type index.

$$I_c = \sqrt{(3.47 - \log Q)^2 + (1.22 + \log F)^2} = \sqrt{(3.47 - \log 30.4)^2 + (1.22 + \log 1.57)^2} = 2.4$$

Because the CPT soil behavior type index is less than 2.6, recalculate using $n = 0.5$.

$$Q = \left(\frac{q_c - \sigma_{vo}}{p_{\text{atm}}}\right)\left(\frac{p_{\text{atm}}}{\sigma'_{vo}}\right)^n = \left(\frac{40{,}000\ \frac{\text{lbf}}{\text{ft}^2} - 1758\ \frac{\text{lbf}}{\text{ft}^2}}{2117\ \frac{\text{lbf}}{\text{ft}^2}}\right)\left(\frac{2117\ \frac{\text{lbf}}{\text{ft}^2}}{1259\ \frac{\text{lbf}}{\text{ft}^2}}\right)^{0.5} = 23.4$$

$$F = 1.57 \quad \text{(unchanged)}$$

$$I_c = \sqrt{(3.47 - \log Q)^2 + (1.22 + \log F)^2} = \sqrt{(3.47 - \log 23.4)^2 + (1.22 + \log 1.57)^2} = 2.5$$

The CPT soil behavior type index remains less than 2.6 and can therefore be used to determine the correction for grain characteristics, K_c, based on the following criteria.

$$K_c = 1.0 \quad [\text{for } I_c \le 1.64]$$

$$K_c = -0.403I_c^4 + 5.58I_c^3 - 21.63I_c^2 + 33.75I_c - 17.88 \quad [\text{for } I_c > 1.64]$$

Because the CPT soil behavior type index is greater than 1.64, use the appropriate equation.

$$\begin{aligned} K_c &= -0.403I_c^4 + 5.58I_c^3 - 21.63I_c^2 + 33.75I_c - 17.88 \quad [\text{for } I_c > 1.64] \\ &= -(0.403)(2.5)^4 + (5.58)(2.5)^3 - (21.63)(2.5)^2 + (33.75)(2.5) - 17.88 \\ &= 2.75 \end{aligned}$$

Calculate the normalizing factor for the tip penetration value, using $n = 0.5$.

$$C_Q = \left(\frac{P_{\text{atm}}}{\sigma'_{vo}}\right)^n = \left(\frac{2117\ \frac{\text{lbf}}{\text{ft}^2}}{1259\ \frac{\text{lbf}}{\text{ft}^2}}\right)^{0.5} = 1.3 \quad (\le 1.7)$$

$$q_{c1\text{N}} = C_Q\left(\frac{q_c}{P_{\text{atm}}}\right) = (1.3)\left(\frac{40{,}000\ \frac{\text{lbf}}{\text{ft}^2}}{2117\ \frac{\text{lbf}}{\text{ft}^2}}\right) = 24.6$$

$$(q_{c1\text{N}})_{\text{CS}} = K_c q_{c1\text{N}} = (2.75)(24.6) = 67.7$$

Using the correct equation, the value for $\text{CRR}_{M=7.5}$ can be calculated.

$$\text{CRR}_{M=7.5} = (93)\left(\frac{(q_{c1\text{N}})_{\text{CS}}}{1000}\right)^3 + 0.08 = (93)\left(\frac{67.7}{1000}\right)^3 + 0.08 = 0.1$$

The answer is (D).

Why Other Options Are Wrong

(A) This solution fails to correct the CPT tip penetration data for grain characteristics ($K_c = 1$).

(B) This solution mistakenly uses the tip penetration value directly in the equation for $\text{CRR}_{M=7.5}$ (for $(q_{c1\text{N}})_{\text{CS}} < 50$) without prior normalization to atmospheric pressure or correction for grain characteristics.

(C) This solution mistakenly uses the tip penetration value directly in the equation for $CRR_{M=7.5}$ (for $50 \leq (q_{c1N})_{CS} < 160$) without prior normalization to atmospheric pressure or correction for grain characteristics.

SOLUTION 4

To determine the cyclic stress ratio for a magnitude 7.5 earthquake ($CSR_{M=7.5}$), calculate the total overburden and effective stress at the midpoint of the submerged sand layer.

$$\begin{aligned}
\sigma_{vo} &= \sigma'_{vo} + u \\
\sigma_{vo} &= \sigma_{v,\text{dry sand}} + \sigma_{v,\text{sub sand}} \\
&= \gamma_d(1 + w_{\text{dry sand}})D_{\text{dry sand}} \\
&\quad + \gamma_d(1 + w_{\text{sub sand}})D_{\text{sub sand}} \\
&= \left(109.7\ \frac{\text{lbf}}{\text{ft}^3}\right)(1+0.06)(20\ \text{ft}) \\
&\quad + \left(109.7\ \frac{\text{lbf}}{\text{ft}^3}\right)(1+0.15)(3\ \text{ft}) \\
&= 2704\ \text{lbf/ft}^2
\end{aligned}$$

$$\begin{aligned}
\sigma'_{vo} &= \sigma_{vo} - \mu \\
&= 2704\ \frac{\text{lbf}}{\text{ft}^2} - \left(62.4\ \frac{\text{lbf}}{\text{ft}^3}\right)(3\ \text{ft}) \\
&= 2517\ \text{lbf/ft}^2
\end{aligned}$$

Calculate the $CSR_{M=7.5}$.

$$\begin{aligned}
\frac{\tau_{h,\text{ave}}}{\sigma'_{vo}} &= CSR_{M=7.5} = 0.65\left(\frac{a_{\max}}{g}\right)\left(\frac{\sigma_{vo}}{\sigma'_{vo}}\right)r_d \\
&= (0.65)\left(\frac{0.35g}{g}\right)\left(\frac{2704\ \frac{\text{lbf}}{\text{ft}^2}}{2517\ \frac{\text{lbf}}{\text{ft}^2}}\right)(0.95) \\
&= 0.23
\end{aligned}$$

In reference to the volumetric strain graph, estimate the volumetric strain based on the values of $(N_1)_{60} = 7$ and $CSR_{M=7.5} = 0.23$.

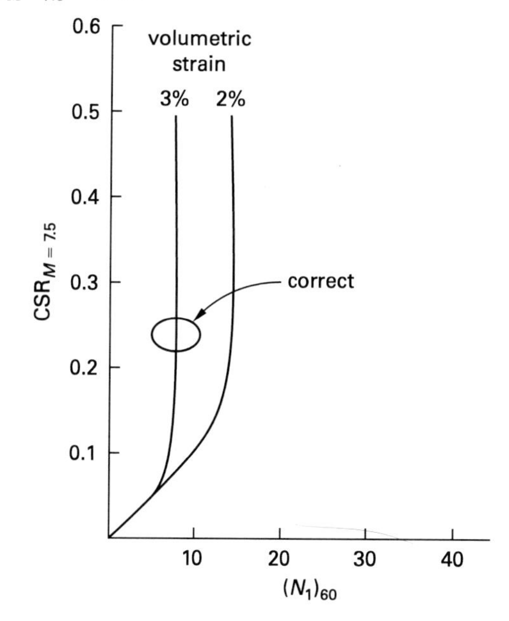

Based on the volumetric strain estimate of 3%, estimate the vertical settlement.

$$\begin{aligned}
S_{M=7.5} &= (\text{volumetric strain})L_{\text{sub sand}} \\
&= (0.03)(6\ \text{ft})\left(12\ \frac{\text{in}}{\text{ft}}\right) \\
&= 2.2\ \text{in}
\end{aligned}$$

The answer is (C).

Why Other Options Are Wrong

(A) This solution fails to convert the settlement value from feet to inches.

(B) This solution mistakenly calculates the $CSR_{M=7.5}$ by using the reciprocal of the effective stress and total stress ratio, and then also miscalculates the SPT value by adding all three values. The incorrect values are used in the chart to obtain the incorrect volumetric strain.

(D) This solution fails to find a settlement value and instead reports the volumetric strain of 3% as a value in inches.

SOLUTION 5

The base shear can be calculated.

$$V = \frac{WS_a}{g} = \frac{(35\ \text{kips})\left(0.43\ \frac{\text{ft}}{\text{sec}^2}\right)}{32.2\ \frac{\text{ft}}{\text{sec}^2}}$$
$$= 0.47\ \text{kips}\quad (470\ \text{lbf})$$

The answer is (B).

Why Other Options Are Wrong

(A) This solution fails to convert to pounds-force units.

(C) This solution mistakenly inverts the numerator and denominator in the calculation.

(D) This solution fails to divide the numerator by the gravity.

SOLUTION 6

The sign should be considered a cantilever beam made up of the three metal posts. The natural period of vibration can be determined after the stiffness for a single metal post is calculated and multiplied by three. For a cantilevered beam, the stiffness required to deflect one of the posts 1 ft laterally can be calculated.

$$k = \frac{3EI}{h^3}$$
$$= \frac{(3)\left(2.8\times 10^6\ \frac{\text{lbf}}{\text{in}^2}\right)(17{,}500\ \text{in}^4)\left(\frac{1\ \text{ft}}{12\ \text{in}}\right)^2}{(85\ \text{ft})^3}$$
$$= 1662\ \text{lbf/ft}$$

For the sign to deflect, all three posts must move together. Therefore, the stiffness should be multiplied by three in the calculation for the natural period of vibration.

$$T = 2\pi\sqrt{\frac{W}{gk}}$$
$$= 2\pi\sqrt{\frac{16{,}000\ \text{lbf}}{\left(32.2\ \frac{\text{ft}}{\text{sec}^2}\right)(3)\left(1662\ \frac{\text{lbf}}{\text{ft}}\right)}}$$
$$= 1.98\ \text{sec}$$

The natural frequency of the structure is the reciprocal of the natural period of vibration.

$$f = \frac{1}{T} = \frac{1}{1.98\ \text{sec}}$$
$$= 0.51\ \text{cycles/sec}\quad (0.50\ \text{Hz})$$

The answer is (B).

Why Other Options Are Wrong

(A) This solution fails to multiply the stiffness by three.

(C) This solution fails to determine the reciprocal of the answer to find the frequency and shows a failure to understand that the units of hertz are cycles per second.

(D) This solution fails to multiply the stiffness by three and also fails to determine the reciprocal of the answer to find the frequency.

SOLUTION 7

Because the seismograph is located 100 km from the epicenter, the Richter magnitude can be directly calculated. A_o is typically taken as 0.001 mm.

$$M = \log_{10}\frac{A}{A_o}$$
$$= \log_{10}\frac{25\ \text{mm}}{0.001\ \text{mm}}$$
$$= 4.4$$

The answer is (D).

Why Other Options Are Wrong

(A) This solution mistakenly calculates the Richter magnitude using the lowest amplitude value of 5 mm.

(B) This solution mistakenly calculates the Richter magnitude using a lower amplitude value of 10 mm.

(C) This solution mistakenly calculates the Richter magnitude using the average amplitude value above 5 mm.

SOLUTION 8

The length of the active fault in kilometers can be calculated from the following approximate correlation.

$$\log_{10} L = 1.02M - 5.77$$

Solve for the length of the active fault.

$$L = \left(10^{((1.02)(3.8)-5.77)}\right)\left(1000\ \frac{\text{m}}{\text{km}}\right)$$
$$= 13\ \text{m} \quad (10\ \text{m})$$

The answer is (C).

Why Other Options Are Wrong

(A) This solution fails to convert the answer to meters.

(B) This solution mistakenly converts the answer to meters by multiplying by 10 instead of 1000.

(D) This solution mistakenly converts the answer to meters by multiplying by 10,000 instead of 1000.

SOLUTION 9

Use ASCE/SEI7 Eq. 20.4-2 to calculate the average standard penetration test (SPT) N-value from the provided test results.

$$\overline{N} = \frac{\sum_{i=1}^{n} d_i}{\sum_{i=1}^{n} \frac{d_i}{N_i}}$$
$$= \frac{15\ \text{ft} + 20\ \text{ft} + 30\ \text{ft} + 10\ \text{ft} + 25\ \text{ft}}{\frac{15\ \text{ft}}{12\ \frac{\text{blows}}{\text{ft}}} + \frac{20\ \text{ft}}{16\ \frac{\text{blows}}{\text{ft}}} + \frac{30\ \text{ft}}{15\ \frac{\text{blows}}{\text{ft}}} + \frac{10\ \text{ft}}{19\ \frac{\text{blows}}{\text{ft}}} + \frac{25\ \text{ft}}{17\ \frac{\text{blows}}{\text{ft}}}}$$
$$= 15.4\ \text{blows/ft}$$

From ASCE/SEI7 Table 20.3-1, an SPT N-value of 15.4 blows/ft corresponds to class D, stiff soil.

The answer is (B).

Why Other Options Are Wrong

(A) This incorrect solution is obtained by summing only the SPT N-values, resulting in an SPT N-value of 79 blows/ft, which corresponds to site class C.

(C) This incorrect solution is obtained by calculating only $\sum_{i=1}^{n} \frac{d_i}{N_i}$, resulting in an SPT N-value of 6.5 blows/ft, which corresponds to site class E.

(D) This incorrect solution selects class F, which requires the presence of a water table, clay, or weak cementation. Since none of these conditions were stated in the problem, the soil is not class F.

5 Earth Structures

PAVEMENT STRUCTURES

PROBLEM 1

For a 15 yr period and constant growth rate of 6%, approximate the total equivalent single-axle loads (ESALs) for the design lane if the average daily ESAL for the entire roadway is 3276. The directional distribution factor is 50%, and the lane distribution factor is 0.90.

(A) 1.5×10^3 ESALs

(B) 8.5×10^6 ESALs

(C) 13×10^6 ESALs

(D) 19×10^6 ESALs

Hint: The average daily ESAL for "all lanes" is given.

PROBLEM 2

A new roadway alignment with flexible pavement components is being analyzed for an overlay design. Given the design parameters, layer thickness, and strength coefficients shown, approximate the effective pavement structural number.

design parameters	value
existing asphalt concrete (AC) thickness	4 in
AC strength coefficient	0.30 per inch
existing recycled AC base course (RAB) thickness	3 in
RAB strength coefficient	0.15 per inch
existing aggregate base course (ABC) thickness	5 in
ABC strength coefficient	0.12 per inch
drainage coefficient (all layers)	1.0
ESALs	219,000
initial serviceability index	4.6
terminal serviceability index	2.0
reliability level	90%
overall standard deviation	0.4
resilient modulus, M_R	8000 lbf/in^2

(A) 1.7

(B) 1.8

(C) 2.3

(D) 2.5

Hint: The effective structural number is calculated based on the existing thickness of the pavement components.

SLOPE STABILITY

PROBLEM 3

A transportation department has limited space for temporarily stockpiling embankment fill at an urban location. The proposed location is within an easement approximately 50 ft wide where 12 ft must remain for a roadway. The angle of internal friction of the borrow source soil is 32°. Given that the borrow source soil is primarily a silty sand, approximate the maximum

Earth Structures

height to which the conical stockpiles may be placed while remaining stable from sloughing.

(A) 12 ft

(B) 16 ft

(C) 24 ft

(D) 30 ft

Hint: For silty sand, the angle of internal friction can be assumed to be approximately equal to the angle of repose.

PROBLEM 4

Given the slope shown, the factor of safety using the Taylor method is most nearly

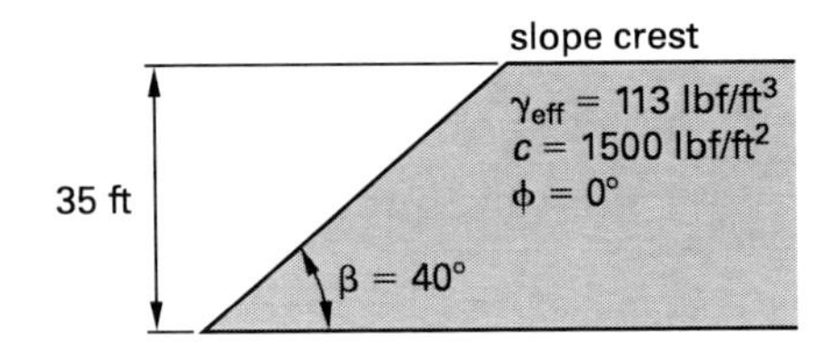

(A) 0.47

(B) 2.1

(C) 2.3

(D) 2.6

Hint: Use the Taylor method with iteration to obtain the factor of safety.

SOLUTION 1

The ESALs for the design lane can be calculated using the equation

$$w_{18} = D_D D_L \widehat{w}_{18}$$

The average daily ESAL for all lanes of the roadway is given as 3276. Therefore, the product of the average annual daily traffic (AADT) and the fraction of AADT that represents truck traffic is 3276 ESALs per day. To determine the total ESALs for the design lane, a growth factor is found from published growth factor tables where the growth rate, g, is read as I.

$$\begin{aligned} F/A, I &= 6\%,\ 15 \text{ yr period} \\ \text{GF} &= 23.3 \end{aligned}$$

Based on the common assumption of 365 days per year, the total ESALs can be determined.

$$\begin{aligned} \widehat{w}_{18} &= (365)(\text{GF for 15 yr period})(\text{daily ESALs}) \\ &= \left(365\ \frac{\text{days}}{\text{yr}}\right)(23.3 \text{ for 15 yr period}) \\ &\quad \times \left(3276\ \frac{\text{ESALs}}{\text{day}}\right) \\ &= 2.79 \times 10^7 \text{ ESALs for 15 yr period} \end{aligned}$$

Calculate the ESALs for the design lane.

$$\begin{aligned} w_{18} &= D_D D_L \widehat{w}_{18} = (0.50)(0.90)(2.79 \times 10^7 \text{ ESALs}) \\ &= 12{,}555{,}000 \text{ ESALs} \quad (13 \times 10^6 \text{ ESALs}) \end{aligned}$$

The answer is (C).

Why Other Options Are Wrong

(A) This option fails to multiply the given daily ESAL value by the growth factor to get the 15 yr total ESAL value. Instead, the given daily ESAL value is multiplied by the design lane information.

(B) This option fails to convert the given growth rate to a monthly value and subsequently calculates the wrong value for the ESALs for the design lane.

(D) This option fails to multiply the total ESAL value for a 15 yr period by the design lane and distribution factors to get the ESALs for the design lane.

SOLUTION 2

Determine the effective structural number using the AASHTO layer-thickness equation. The effective structural number represents the structural number for the existing pavement component layers. Because there are three layers given, three structural coefficients and three drainage coefficients are required along with three thickness values.

$$SN_{eff} = a_1 m_1 D_1 + a_2 m_2 D_2 + a_3 m_3 D_3$$

For the AC layer,

$$\begin{aligned} a_1 &= \frac{0.30}{1 \text{ in}} \\ m_1 &= 1.0 \\ D_1 &= 4 \text{ in} \end{aligned}$$

For the RAB layer,

$$\begin{aligned} a_2 &= \frac{0.15}{1 \text{ in}} \\ m_2 &= 1.0 \\ D_2 &= 3 \text{ in} \end{aligned}$$

For the ABC layer,

$$\begin{aligned} a_3 &= \frac{0.12}{1 \text{ in}} \\ m_3 &= 1.0 \\ D_3 &= 5 \text{ in} \end{aligned}$$

The effective structural number can be calculated.

$$\begin{aligned} SN_{eff} &= a_1 m_1 D_1 + a_2 m_2 D_2 + a_3 m_3 D_3 \\ &= \left(\frac{0.30}{1 \text{ in}}\right)(1.0)(4 \text{ in}) + \left(\frac{0.15}{1 \text{ in}}\right)(1.0)(3 \text{ in}) \\ &\quad + \left(\frac{0.12}{1 \text{ in}}\right)(1.0)(5 \text{ in}) \\ &= 2.25 \quad (2.3) \end{aligned}$$

The answer is (C).

Why Other Options Are Wrong

(A) This option fails to include the ABC layer in the calculation.

(B) This option fails to include the RAB layer in the calculation.

(D) This option mistakenly calculates a new overall design structural number using the AASHTO nomograph for flexural pavement design, $SN_{design} = 2.5$.

SOLUTION 3

Assuming the angle of internal friction is equal to the angle of repose for silty sand, the maximum slope angle of the stockpiles can be considered equivalent to the angle of repose. Using the following relationship, the height of the stockpiles can be determined by

$$\frac{H}{r} = \tan\beta$$

It is reasonable to assume that the angle of repose is approximately equal to the angle of internal friction, since the borrow source soil consists of a silty sand, not a clean, poorly graded sand. Therefore, it is conservative to assume that the maximum and stable slope angle will be equal to or less than the angle of internal friction. The maximum stockpile diameter that can fit the available width is

$$\begin{aligned} d &= 2r = w_{easement} - w_{roadway} \\ &= 50 \text{ ft} - 12 \text{ ft} \\ &= 38 \text{ ft} \end{aligned}$$

Therefore, the maximum allowable radius is

$$\begin{aligned} r &= \frac{d}{2} = \frac{38 \text{ ft}}{2} \\ &= 19 \text{ ft} \end{aligned}$$

The maximum height of the stockpile slope can be found.

$$\begin{aligned} H &= r \tan\beta \\ &= (19 \text{ ft})\tan 32^\circ \\ &= 12 \text{ ft} \end{aligned}$$

The answer is (A).

Why Other Options Are Wrong

(B) This incorrect solution does not allow for the 12 ft roadway. The height is greater than the correct answer.

(C) This incorrect solution is obtained by using the diameter of the stockpile instead of the radius. The height is double the amount of the correct answer.

(D) This incorrect solution is obtained by using the reciprocal of the correct ratio and proceeding to find the incorrect maximum height. The answer obtained is much greater than the correct answer.

SOLUTION 4

Determine the factor of safety with respect to cohesion using the Taylor slope stability chart for the angle of internal friction equivalent to zero. Because a depth to a firm base beneath the slope is not given, assume the depth, D, is infinite, and determine the stability number, N_o.

$$d = \infty$$
$$\beta = 40°$$
$$N_o = 5.53$$

Calculate the corresponding factor of safety with respect to cohesion.

$$F_{\text{cohesive}} = \frac{N_o c}{\gamma_{\text{eff}} H} = \frac{(5.53)\left(1500\ \frac{\text{lbf}}{\text{ft}^2}\right)}{\left(113\ \frac{\text{lbf}}{\text{ft}^3}\right)(35\ \text{ft})} = 2.1$$

The answer is (B).

Why Other Options Are Wrong

(A) This solution mistakenly inverts the stability number equation.

(C) This solution misinterprets the Taylor slope stability chart and misreads the stability number as 6.

(D) Because a depth value to a firm base is not given, it is possible for this incorrect solution to be obtained by assuming the depth factor, d, is equivalent to zero and arriving at a stability number of 6.75.

Groundwater and Seepage

SEEPAGE ANALYSIS/GROUNDWATER FLOW

PROBLEM 1

Steady-state vertical seepage is occurring in the soil profile shown.

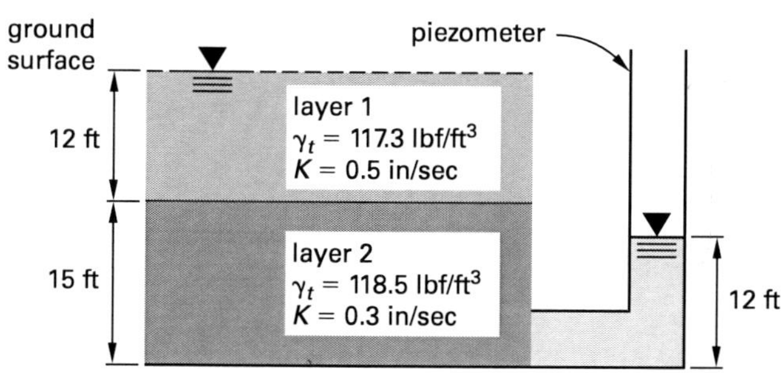

Approximate the effective stress at the base of the upper soil layer.

(A) 470 lbf/ft^2

(B) 660 lbf/ft^2

(C) 900 lbf/ft^2

(D) 970 lbf/ft^2

Hint: Use Darcy's law to equate the volumetric flow rate for both layers.

PROBLEM 2

For the 50 ft wide sheet pile dam shown, approximate the seepage rate. The coefficient of permeability for the soil is 3×10^{-5} in/sec.

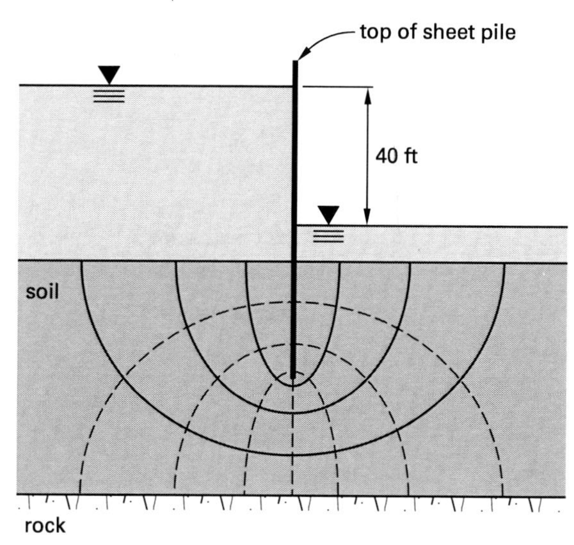

(A) 5.0×10^{-5} ft^3/sec

(B) 2.5×10^{-3} ft^3/sec

(C) 5.0×10^{-3} ft^3/sec

(D) 3.0×10^{-2} ft^3/sec

Hint: Use Darcy's law along with the number of flow channels and equipotential drops to determine the seepage rate.

PROBLEM 3

A homogeneous earth dam and flow net are shown. The coefficient of permeability, K, is 3.7×10^{-4} ft/min.

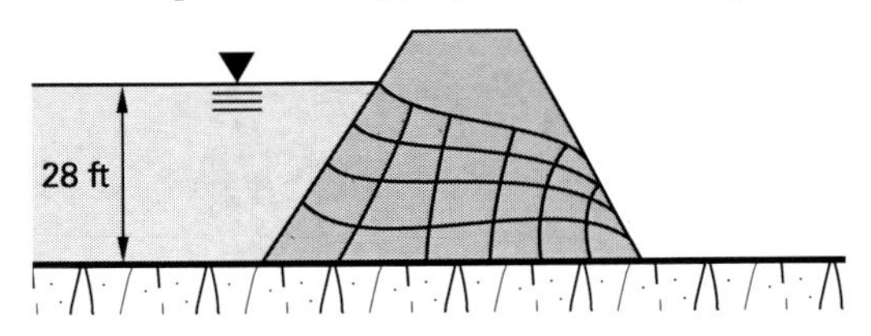

What is most nearly the approximate seepage rate per unit width of dam if the water height behind the dam is 28 ft?

(A) 8.6×10^{-5} ft^3/sec

(B) 12×10^{-5} ft^3/sec

(C) 1.7×10^{-4} ft^3/sec

(D) 6.9×10^{-3} ft^3/sec

Hint: The given flow net has four flow paths and six equipotential drops.

SOLUTION 1

Because hydrostatic conditions do not exist, Darcy's law can be used to equate the volumetric flow rate for both soil layers.

$$Q = KiA$$

The volumetric flow rate through both soil layers is equivalent, and a constant unit value for the flow area can be assumed.

$$Q_1 = Q_2 = (Ki)_1 A = (Ki)_2 A$$
$$(Ki)_1 = (Ki)_2$$

The hydraulic gradient for both soil layers can be expressed.

$$i = \frac{\Delta h}{L}$$
$$i_1 = \frac{\Delta h_1}{12 \text{ ft}}$$
$$i_2 = \frac{\Delta h_2}{15 \text{ ft}}$$

Use Darcy's law to equate the volumetric flow rates and determine the change in head through each soil layer. The units for the coefficient of permeability divide out.

$$(Ki)_1 = (Ki)_2$$
$$\left(0.5 \ \frac{\text{in}}{\text{sec}}\right)\left(\frac{\Delta h_1}{12 \text{ ft}}\right) = \left(0.3 \ \frac{\text{in}}{\text{sec}}\right)\left(\frac{\Delta h_2}{15 \text{ ft}}\right)$$
$$\Delta h_1 = (0.48)\Delta h_2$$

The sum of the change in head for both layers is equivalent to the total head loss measured through both soil layers.

$$\Delta h_1 + \Delta h_2 = 15 \text{ ft}$$
$$(0.48)\Delta h_2 + \Delta h_2 = 15 \text{ ft}$$
$$\Delta h_2 = 10.1 \text{ ft}$$
$$\Delta h_1 = 4.9 \text{ ft}$$

The pore pressure due to steady-state seepage conditions at the base of the upper soil layer is equivalent to the piezometric head measured at the base of the upper soil layer. The piezometric head at the base of the upper soil layer can be calculated by subtracting the change in head through the upper soil layer from the total piezometric head measured at the base of the lower soil layer.

$$\begin{aligned} h_1 &= 12 \text{ ft} - \Delta h_1 = 12 \text{ ft} - 4.9 \text{ ft} \\ &= 7.1 \text{ ft} \\ u &= (7.1 \text{ ft})\left(62.4 \ \frac{\text{lbf}}{\text{ft}^3}\right) \\ &= 443 \text{ lbf/ft}^2 \end{aligned}$$

Calculate the total and effective stresses.

$$\begin{aligned} \sigma_1 &= \gamma_{t,1} D_1 = \left(117.3 \ \frac{\text{lbf}}{\text{ft}^3}\right)(12 \text{ ft}) \\ &= 1408 \text{ lbf/ft}^2 \\ \sigma' &= \sigma_1 - u = 1408 \ \frac{\text{lbf}}{\text{ft}^2} - 443 \ \frac{\text{lbf}}{\text{ft}^2} \\ &= 965 \text{ lbf/ft}^2 \quad (970 \text{ lbf/ft}^2) \end{aligned}$$

The answer is (D).

Why Other Options Are Wrong

(A) This incorrect answer is obtained by assuming that the pore pressure is equivalent to the hydrostatic head at the base of the lower soil layer less the piezometric head at that level.

(B) This incorrect answer is obtained by assuming the pore pressure is equivalent to the hydrostatic head at the base of the upper soil layer.

(C) This incorrect solution is obtained by using the wrong value for the change in head through the upper soil layer.

SOLUTION 2

The seepage rate can be determined from Darcy's law.

$$Q = KHL\left(\frac{N_f}{N_d}\right)$$

Count the number of flow channels and equipotential drops (respectively).

$$\begin{aligned} N_f &= 4 \\ N_d &= 8 \end{aligned}$$

The seepage rate is

$$\begin{aligned} Q &= KHL\left(\frac{N_f}{N_d}\right) \\ &= \left(3 \times 10^{-5} \ \frac{\text{in}}{\text{sec}}\right)\left(\frac{1 \text{ ft}}{12 \text{ in}}\right)(40 \text{ ft})(50 \text{ ft})\left(\frac{4}{8}\right) \\ &= 2.5 \times 10^{-3} \text{ ft}^3/\text{sec} \end{aligned}$$

The answer is (B).

Why Other Options Are Wrong

(A) Failure to multiply the resulting seepage by the length of the dam will result in this answer. This result would have been acceptable if the problem had requested the seepage observed per foot of dam length.

(C) This incorrect solution does not include conversion of the coefficient of permeability to units of feet. The units do not work out correctly, and the value is greater than the correct value.

(D) If the number of flow channels is not multiplied and the number of equipotential drops is not divided, this incorrect solution results.

SOLUTION 3

In reference to the given flow net, the rate of seepage per unit width of the earth dam can be determined when the height of the water is 28 ft behind the dam.

$$\begin{aligned} Q &= kH\left(\frac{N_f}{N_p}\right) \\ &= \left(\frac{3.7 \times 10^{-4} \ \frac{\text{ft}}{\text{min}}}{60 \ \frac{\text{sec}}{\text{min}}}\right)(28 \text{ ft})\left(\frac{4}{6}\right) \\ &= 12 \times 10^{-5} \text{ ft}^3/\text{sec per ft of dam width} \end{aligned}$$

The answer is (B).

Why Other Options Are Wrong

(A) This solution is incorrect because it uses three flow channels instead of four.

(C) This solution neglects to use the given flow net.

(D) This solution mistakenly omits the conversion of the coefficient of permeability to feet per second.

7 Problematic Soil and Rock Conditions

EXPANSIVE AND SENSITIVE SOILS

PROBLEM 1

Based on a preliminary pilot boring conducted at the location of a pier foundation, the subsurface soils are expected to be soft saturated clay with layers of silt and fine sand. What type of sampling and testing equipment is best suited for obtaining reliable in-situ engineering properties for design of driven precast concrete piles?

(A) cone penetration test (CPT) apparatus

(B) hand auger in test pit excavation

(C) split-spoon standard penetration test (SPT) sampler with blow counts

(D) thin-wall tube sampler

Hint: The goal of the exploration program is to obtain relatively undisturbed samples for laboratory testing and development of design parameters best suited for a driven pile foundation in soft saturated clay.

PROBLEM 2

A bridge is proposed for a stream crossing with intermittent seasonal flow and is located in a zone of negligible seismic activity. The proposed foundation plan consists of two abutments to be constructed behind the banks away from the bed of the stream and a single middle pier to be supported entirely by precast concrete piles driven into the subsurface soils at the center of the stream. Based on a preliminary pilot boring conducted at the location of the center pier, the subsurface soils are expected to be soft saturated clay with silt and layers of fine sand. To reliably evaluate the vertical load capacity of the piles, list the required engineering properties and soil profile characteristics to be obtained from a field exploration at the site.

I. Atterberg limits, gradation, and standard penetration test (blow counts)

II. cohesion, shear strength, angle of internal friction, and soil-pile friction angle

III. compression index, coefficient of vertical consolidation, and permeability

IV. in-situ density, strata-layer thickness, and groundwater table depth

(A) I and II

(B) II and III

(C) II and IV

(D) I, III, and IV

Hint: Consider the equations required to evaluate vertical pile capacity, keeping in mind the assumption that the entire load at the center of the stream will be placed on piles and not on the adjacent ground.

PROBLEM 3

A project site for a multistory hotel is underlain by sandy clay that is characterized by moderate swell potential extending to a depth of 30 ft below the ground surface as shown. Below 30 ft, the clay is underlain by dense sand with gravel to a depth of 45 ft below the ground surface.

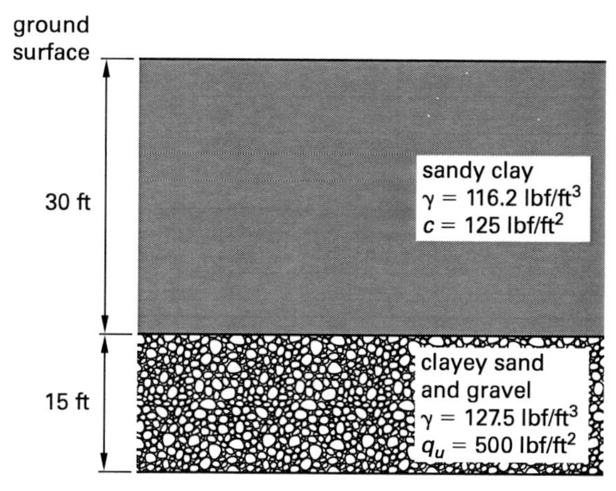

If 30 in diameter cast-in-place concrete drilled piers are being designed for a 20 kip allowable capacity (factor of

Problematic Soil & Rock Conds.

safety of 2), use average values from the α-method to approximate the minimum required embedment length.

(A) 29 ft

(B) 30 ft

(C) 35 ft

(D) 41 ft

Hint: Use the factor of safety to determine the required drilled pier length for the ultimate capacity. Assume the absence of end bearing capacity to find the greatest pier length needed for the loading given.

SOLUTION 1

A thin-wall tube sampler is most appropriate for development of foundation design parameters needed for driven piles in the soft saturated clay expected at the site. Thin-wall tube samplers are a superior choice among the other options because they can obtain fairly undisturbed samples for triaxial shear strength testing. A thin-wall tube sampler is also capable of obtaining sufficient sample quantity for purposes of soil classification and moisture content determination for development of the subsurface soil profile.

The answer is (D).

Why Other Options Are Wrong

(A) This answer is incorrect. A CPT apparatus provides reliable results for frictional soil resistance and end resistance; however, soil samples are not recoverable for visual observation and analysis. The skin-friction and end-resistance data may be used with reliability for driven pile design and liquefaction analysis (where required), but the apparatus does not permit evaluation of moisture content and soil classification using standardized test methods for subsequent development of a soil profile.

(B) This answer is incorrect. Excavation of a test pit to sufficient depth would be difficult given the expected presence of groundwater. The likelihood of soil disturbance is also very high compared to drilling and sampling with a thin-wall tube sampler. Excessive soil disturbance results in unsuitable samples for determination of triaxial strength test values and laboratory properties such as in-situ density and volume.

(C) This answer is incorrect. A split-spoon sampler may be used to obtain SPT blow counts; however, data correlation in reference to researched information would be required to develop the engineering properties needed to design the driven pile foundation. A thin-wall tube sampler would allow sampling of the actual soils with minimal disturbance. Although soil samples of sufficient quantity can be obtained with a split-spoon sampler for purposes of soil classification and moisture content determination, the samples would typically be more disturbed than thin-wall tube samples, making them unsuitable for triaxial shear strength testing.

SOLUTION 2

To evaluate vertical pile capacity, the properties listed under groups II and IV are required by equations to calculate skin friction and end bearing.

The answer is (C).

Why Other Options Are Wrong

(A) The properties listed in this choice constitute only a portion of the parameters required to calculate vertical pile capacity from skin friction and end bearing.

(B) Although this choice includes a portion of the parameters required to calculate vertical pile capacity from skin friction and end bearing, the remaining parameters—such as density, strata-layer thickness, and location of groundwater for determination of effective stress—are not included. The remaining parameters included in this choice are appropriate for calculation of consolidation settlement and would be needed if loading at the ground surface near the proposed pile was planned.

(D) The properties listed in this choice constitute only a portion of the parameters required to calculate vertical pile capacity from skin friction and end bearing. Properties such as density, strata-layer thickness, and location of groundwater for determination of effective stress are necessary. However, the remaining parameters are appropriate for soil classification and calculation of consolidation settlement rather than for vertical pile capacity. Although strength correlation based on the other parameters may exist for determination of vertical pile capacity, this methodology is not the most reliable and therefore is not the most desirable.

SOLUTION 3

The drilled pier length can be determined by trial and error, by assuming the absence of end bearing capacity to find the greatest pier length required for the loading.

$$\begin{aligned} Q_a &= \frac{Q_{\text{ult}}}{2} \\ Q_{\text{ult}} &= Q_f + Q_p = 2Q_a \\ &= (2)(20 \text{ kips}) \\ &= 40 \text{ kips} \\ Q_f &= f_s \pi B L = 40 \text{ kips} \\ Q_p &= N_c c A_p \end{aligned}$$

The α-method is requested for the friction capacity determination. Solve the friction capacity equation for pier embedment length, determine the value for α from published tables, and use the cohesion value for the upper soil layer.

$$\begin{aligned} f_s &= \alpha c \\ L &= \frac{Q_f}{f_s \pi B} = \frac{Q_f}{\alpha c \pi B} \\ &= \frac{40 \text{ kips}}{(1)\left(0.125 \frac{\text{kips}}{\text{ft}^2}\right)\pi(30 \text{ in})\left(\frac{1 \text{ ft}}{12 \text{ in}}\right)} \\ &= 41 \text{ ft} \end{aligned}$$

The upper soil layer extends only to a depth of 30 ft. Therefore, 41 ft worth of friction capacity is not possible using properties from the upper soil layer because the friction capacity has to be limited to 30 ft. Before taking the time to determine the friction capacity for the portion of the pier extending 11 ft into the lower soil layer, check the friction capacity for a 30 ft deep embedment.

$$\begin{aligned} Q_f &= \alpha c \pi B L \\ &= (1)\left(0.125 \frac{\text{kip}}{\text{ft}^2}\right)\pi(30 \text{ in}) \\ &\quad \times \left(\frac{1 \text{ ft}}{12 \text{ in}}\right)(30 \text{ ft}) \\ &= 29.5 \text{ kips} \end{aligned}$$

Determine the tip capacity for end bearing on the underlying clayey sand and gravel layer and add it to the friction capacity to find if adequate allowable capacity is available. N_c is acceptable to be taken as 9.

$$\begin{aligned} Q_p &= N_c\left(\frac{q_u}{2}\right)A_p \\ &= (9)\left(\frac{0.500 \frac{\text{kips}}{\text{ft}^2}}{2}\right)\left(\frac{\pi}{4}\right)(30 \text{ in})^2\left(\frac{1 \text{ ft}}{12 \text{ in}}\right)^2 \\ &= 11 \text{ kips} \end{aligned}$$

$$\begin{aligned} Q_{\text{ult}} &= Q_f + Q_p = 29.5 \text{ kips} + 11 \text{ kips} \\ &= 40.5 \text{ kips} \\ Q_a &= \frac{Q_{\text{ult}}}{2} = \frac{40.5 \text{ kips}}{2} \\ &= 20.3 \text{ kips} \end{aligned}$$

A 30 ft deep embedment is acceptable.

The answer is (B).

Why Other Options Are Wrong

(A) This option is incorrect because 29 ft of embedment is not enough to achieve adequate friction capacity and permit end bearing in the stronger clayey sand and gravel soil.

(C) This option is incorrect because it ignores the end bearing capacity and obtains the remaining needed friction capacity from the lower layer. The required length would be 35 ft.

(D) This option is incorrect because the minimum length for the pier is requested. Although it is possible to obtain 41 ft worth of friction capacity from both soil layers, the allowable capacity can be obtained from an embedment of 30 ft with end bearing on the clayey sand and gravel layer.

Earth Retaining Structures

FLEXIBLE RETAINING WALL STABILITY

PROBLEM 1

An anchored sheet pile will be installed during an excavation and keyed into the sandstone bedrock to allow rotation, but it will be horizontally restricted. A deadman anchor will be installed behind the active zone, and a tie rod will be extended to the sheet pile at a depth of 8 ft below the ground surface.

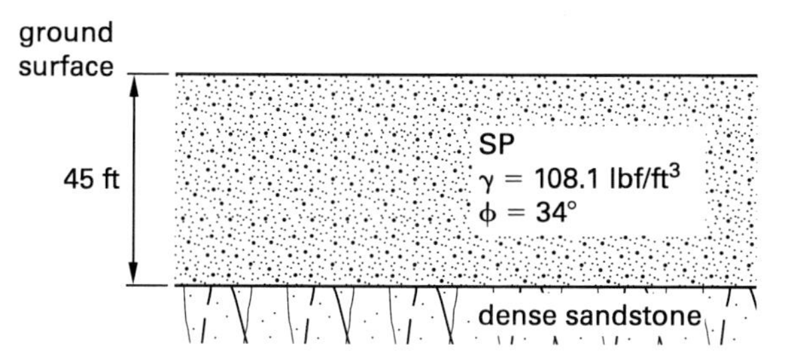

Assuming no contribution from passive pressure from within the sandstone, approximate the tensile force in the tie rod per unit length of wall.

(A) 10 kips/ft

(B) 12 kips/ft

(C) 25 kips/ft

(D) 37 kips/ft

Hint: Because the sheet pile face is vertical and the backfill is horizontal, the Rankine equation can be used.

LATERAL EARTH PRESSURE

PROBLEM 2

The cantilevered retaining wall shown is under design.

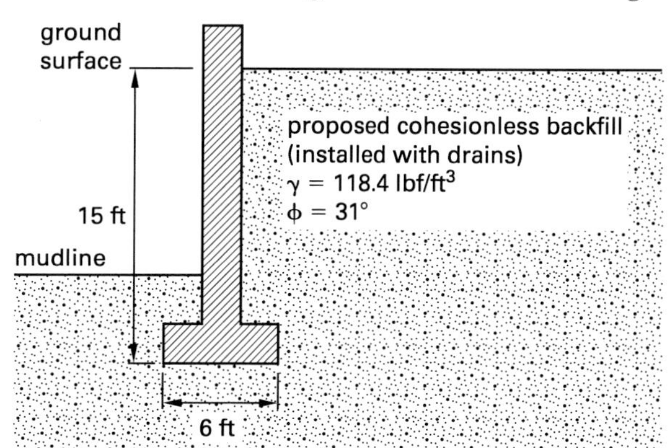

What is most nearly the active resultant per unit length of wall?

(A) 280 lbf/ft

(B) 1900 lbf/ft

(C) 4300 lbf/ft

(D) 8500 lbf/ft

Hint: Calculate the coefficient of active earth pressure, and use it to find the active resultant.

PROBLEM 3

Given the retaining wall shown, use the Coulomb equation to approximate the active resultant.

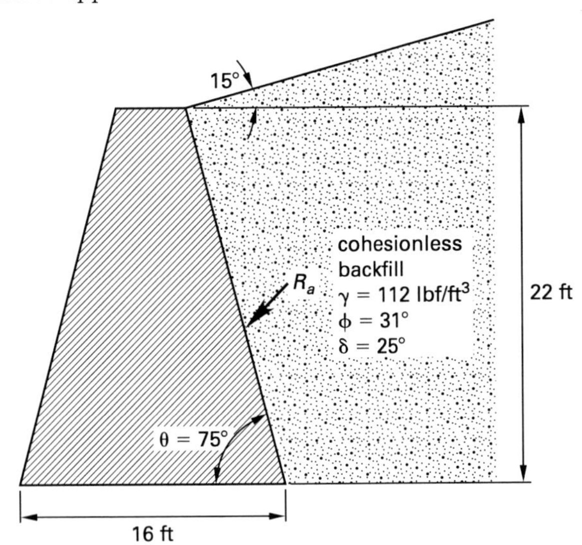

(A) 8680 lbf/ft

(B) 14,700 lbf/ft

(C) 19,200 lbf/ft

(D) 29,300 lbf/ft

Hint: The Coulomb equation allows for sloping backfill, inclined backfill face, and friction between the soil and the retaining wall.

PROBLEM 4

The cantilevered retaining wall shown is under design.

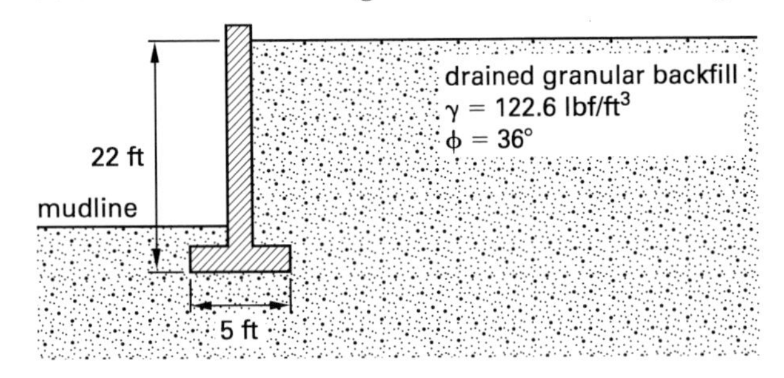

What is most nearly the active resultant per unit length of wall?

(A) 350 lbf/ft

(B) 3400 lbf/ft

(C) 7700 lbf/ft

(D) 15,000 lbf/ft

Hint: Calculate the coefficient of active earth pressure, and use it to find the active resultant.

PROBLEM 5

An excavation is planned to a depth of 51 ft and is currently two-thirds complete in the soil profile shown.

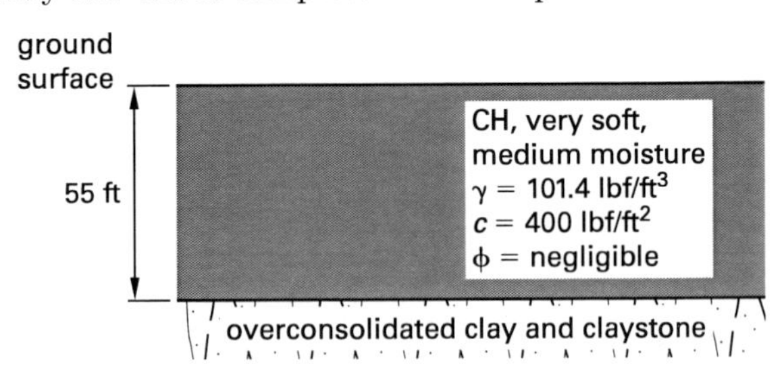

Approximate the maximum lateral earth pressure existing at the midpoint of the current excavation depth.

(A) 120 lbf/ft²

(B) 990 lbf/ft²

(C) 1850 lbf/ft²

(D) 3600 lbf/ft²

Hint: Calculate the stability number to determine the lateral earth pressure diagram to be used.

RIGID RETAINING WALL STABILITY

PROBLEM 6

The gravity wall shown has been designed to resist overturning and sliding. The total vertical force component made up of the self-weight of the wall is 14 kips per unit length of wall.

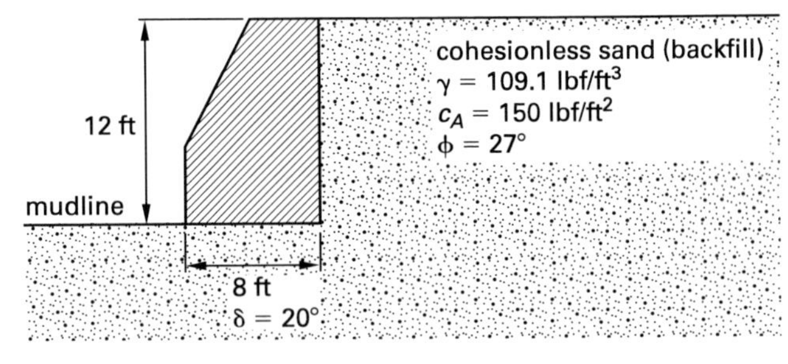

The approximate factor of safety against sliding is

(A) 1.1

(B) 1.6

(C) 1.7

(D) 2.1

Hint: The active pressure of the backfill is resisted by friction between the base of the wall and the soil.

PROBLEM 7

A cantilevered retaining wall was originally designed with a base width of 10 ft, but it was actually constructed with a base width of 6 ft. The original design incorporated a factor of safety of 2 and ignored passive pressure. Given the backfill conditions shown, determine the new factor of safety against overturning moment about the toe.

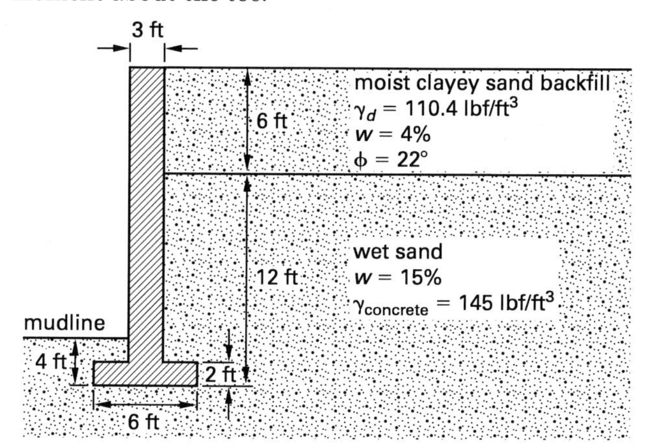

(A) 0.68

(B) 0.80

(C) 1.2

(D) 1.4

Hint: Determine the active resultants, overturning moment, and resisting moment.

PROBLEM 8

Given the height of the backfill shown, what is most nearly the height of the keyless concrete gravity wall shown required to adequately resist sliding with a factor of safety of 2? Use 150 lbf/ft^3 for the concrete unit weight and ignore cohesion.

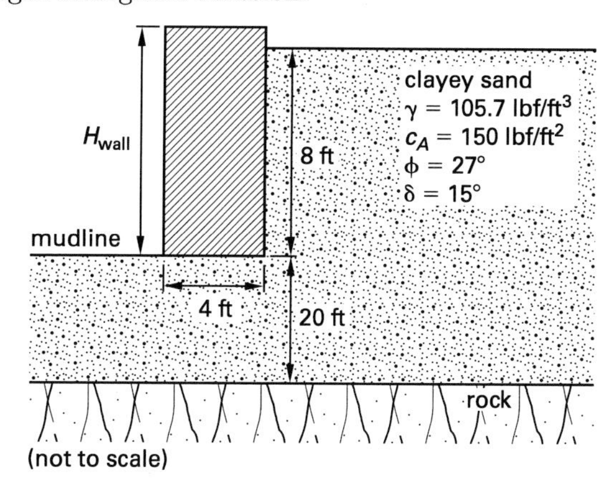

(A) 6.0 ft

(B) 12 ft

(C) 21 ft

(D) 28 ft

Hint: Based on the height of the wall, the weight of the wall is determined from the unit weight of concrete and the volume of the wall.

PROBLEM 9

After several years of service, the retaining wall shown was found to have plugged drains and 7 ft of compound water above the footing level. After performing a forensic field exploration, laboratory testing was conducted on the backfill soil. The difference in strength parameters was found to be negligible in comparison to the original design conditions.

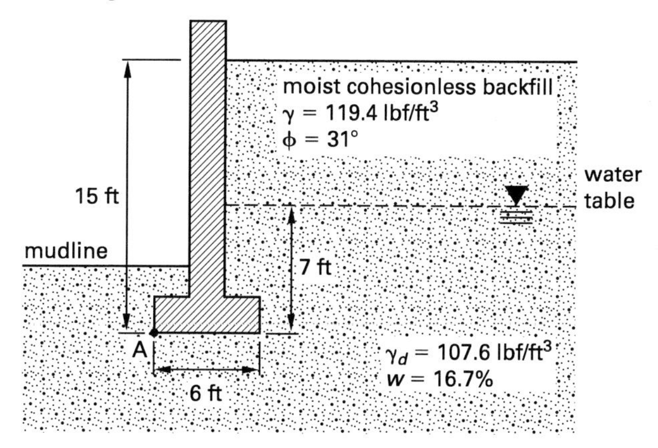

Earth Retaining Structures

The resulting overturning moment, per unit length of wall, about point A is

(A) 21 ft-kips/ft

(B) 24 ft-kips/ft

(C) 37 ft-kips/ft

(D) 48 ft-kips/ft

Hint: Determine the active resultant for the soil both above and below the phreatic surface by separate calculations.

PROBLEM 10

Three years after construction of the retaining wall shown, a parking area with a pavement of negligible thickness is proposed. The paved parking area is approximately 20 ac in size and is bordered on one side by the retaining wall.

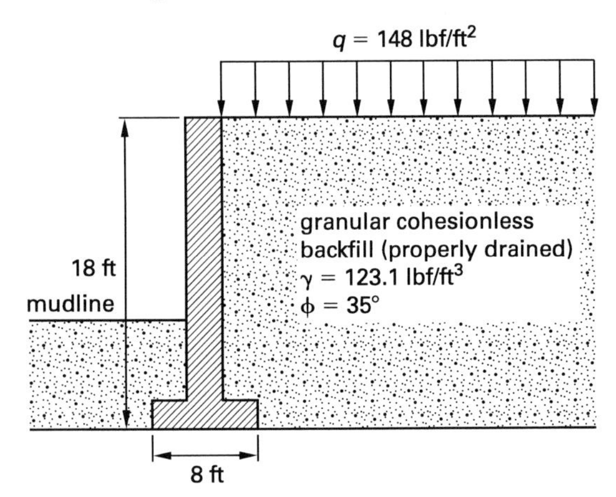

Approximate the overturning moment about the toe.

(A) 8.3 ft-kips

(B) 21 ft-kips

(C) 39 ft-kips

(D) 56 ft-kips

Hint: Determine the existing overturning moment and then treat the surcharge pressure as a uniform pressure distribution acting horizontally along the entire height of the backfill face of the wall and add that resulting additional moment.

PROBLEM 11

Approximate the resistance to sliding for the keyed gravity wall shown where the key is located a negligible distance from the heel. The total vertical force component made up of the self-weight of the wall is 16.5 kips/ft of wall length.

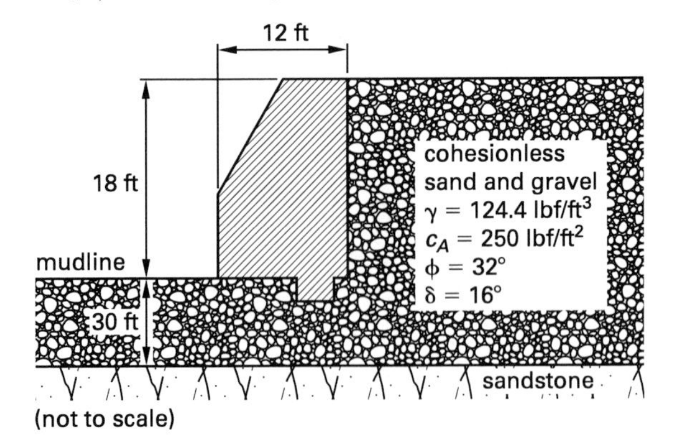

(A) 7.7 kips/ft

(B) 10 kips/ft

(C) 11 kips/ft

(D) 13 kips/ft

Hint: The active pressure of the backfill is resisted by the internal friction (shear strength) of the soil located in front of the key, as well as friction (adhesion) between the soil and base of the wall.

PROBLEM 12

The cantilevered retaining wall shown was originally designed with a base width of 6 ft, but it was actually constructed with a base width of 4 ft. The original design incorporated a factor of safety of 2 and ignored passive pressure.

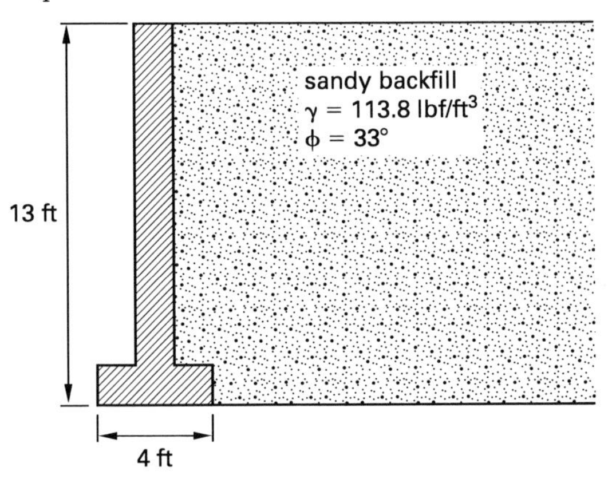

Approximate the overturning moment about the toe.

(A) 3.2 ft-kips

(B) 12 ft-kips

(C) 18 ft-kips

(D) 36 ft-kips

Hint: Determine the active resultant and overturning moment.

PROBLEM 13

The cantilevered retaining wall shown is being constructed along the property line at the edge of a golf course.

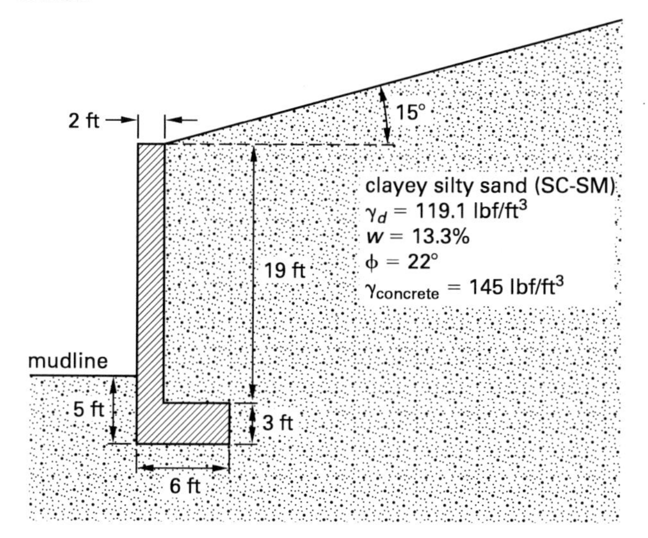

Ignoring the passive earth pressure, what is most nearly the approximate factor of safety against overturning?

(A) 0.13

(B) 0.78

(C) 2.2

(D) 6.5

Hint: Use Coulomb theory to calculate the overturning and resisting moments to find the factor of safety.

PROBLEM 14

The soil profile shown existed prior to construction of the retaining wall and backfill shown. The elevation of the original ground surface equals the top elevation of the retaining wall backfill.

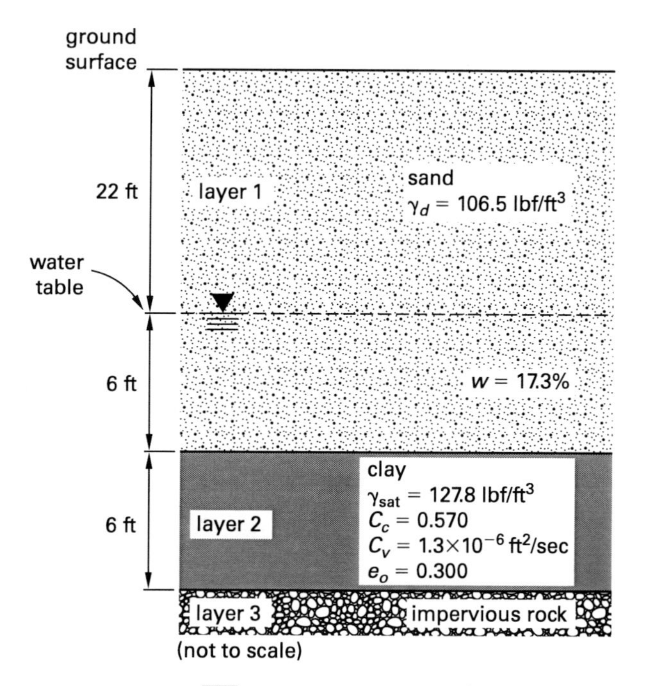

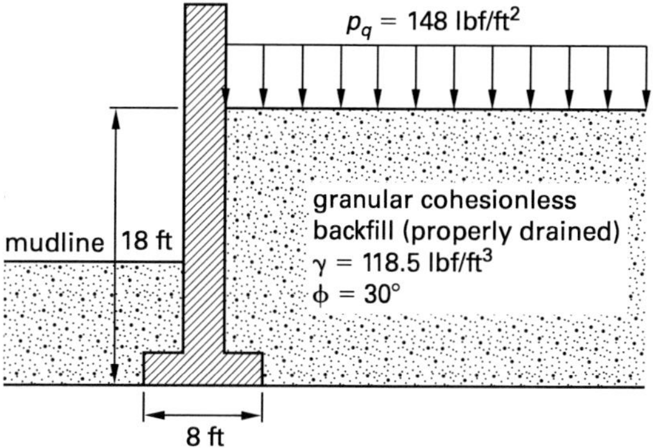

Ignoring the retaining wall self-weight, approximate the settlement of the backfill.

(A) 0 in

(B) 2 in

(C) 3 in

(D) 8 in

Hint: The retaining wall, backfill, and surcharge may increase the effective stress in the clay layer.

TIE-BACKS

PROBLEM 15

The mechanically reinforced retaining wall shown is under design.

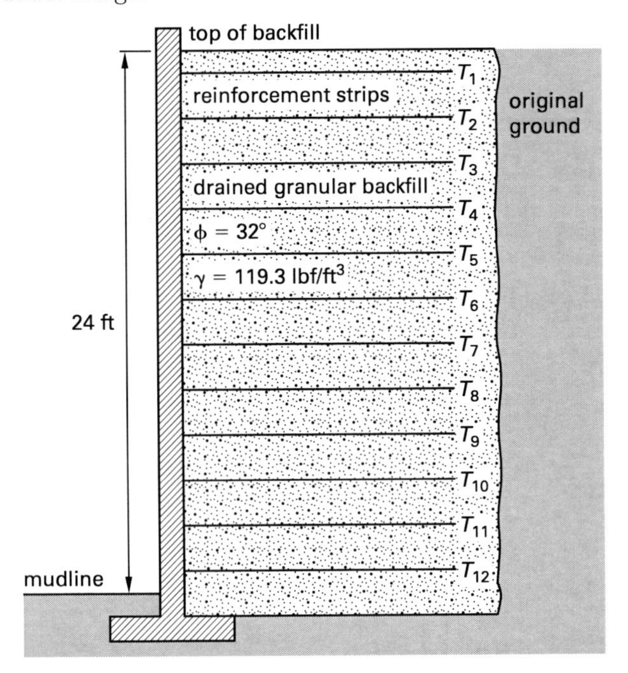

Rectangular shaped reinforcing strips are to be spaced at 2 ft horizontal intervals on center and at vertical heights that are 2 ft apart. Given a strip width and thickness of 3 in and 0.25 in, respectively, what is most nearly the approximate length of the reinforcing strips if the friction angle between the metal strips and the backfill soil, δ, is 20°? Ignore passive resistance and the strip thickness contribution to frictional resistance.

(A) 7.0 ft

(B) 17 ft

(C) 20 ft

(D) 27 ft

Hint: Develop a table to calculate the tension development, available friction resistance, and length of each strip. Use the following equation for calculating the tension value of each strip: $T_i = \gamma d_i h s k_a$.

PROBLEM 16

The mechanically reinforced retaining wall shown will be constructed using thin metal rods for reinforcement. The rods are to be spaced at 3 ft horizontal intervals on center and at vertical intervals 2 ft apart. Ignore passive resistance.

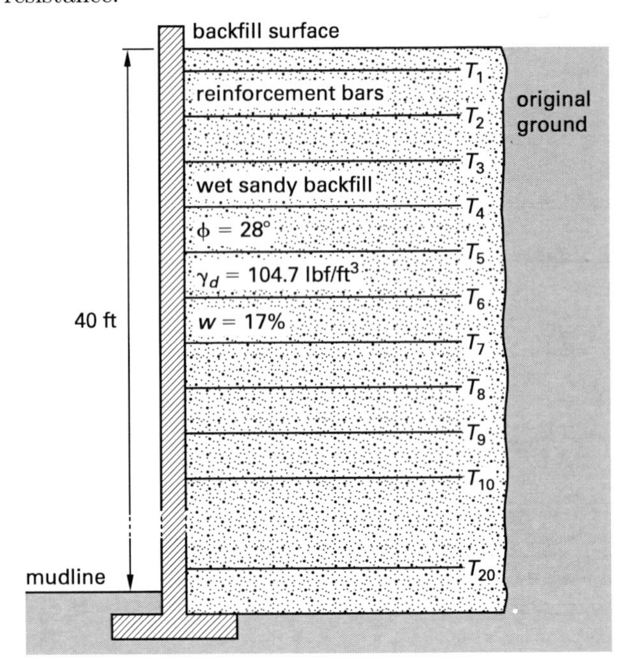

Given a rod diameter of 2 in, what is most nearly the tension required in reinforcing rod no. 8 (from the top) if the friction angle between the metal rods and the backfill soil, δ, is 18°?

(A) 0.50 kips

(B) 1.3 kips

(C) 4.0 kips

(D) 31 kips

Hint: Calculate the tension required using the coefficient of active earth pressure.

Earth Retaining Structures

PROBLEM 17

An anchored flexible bulkhead, fixed from rotation and horizontal movement, will be used to support the excavation walls in a large foundation construction project. The bulkhead will be installed to a total depth of 35 ft, and the excavation will be cut to a depth of 20 ft in the soil profile shown. The tie rod will be anchored at a sufficient distance behind the active wedge at a depth of 5 ft below the ground surface.

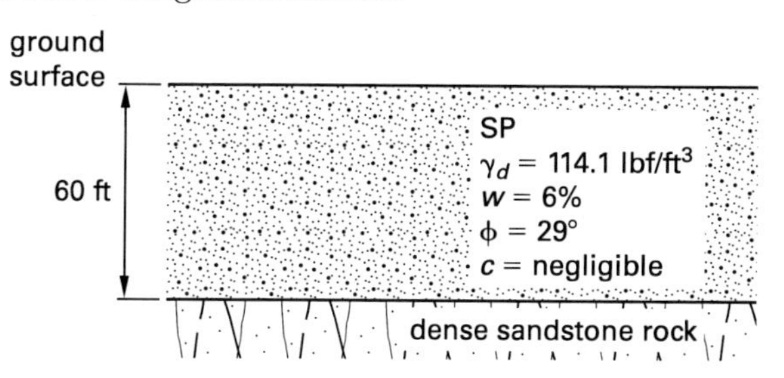

What is most nearly the approximate tensile force in the tie rod per unit width of bulkhead?

(A) 4.9 kips

(B) 5.8 kips

(C) 6.1 kips

(D) 15 kips

Hint: Because the bulkhead is anchored and fixed from rotation, there will be a point of inflection (or pivot point) below the cut line where the moment will be zero.

SOLUTION 1

The equation for active horizontal earth pressure can be used to find the active earth pressure acting at the base of the sand layer (SP, poorly graded sand). An equation for passive horizontal earth pressure is not necessary.

$$p_a = p_v k_a$$

The Rankine equation for the active earth pressure coefficient may be used.

$$k_a = \frac{1 - \sin\phi}{1 + \sin\phi}$$

Calculate the active horizontal earth pressure at the base of the sand, starting with the active earth pressure coefficient.

$$\begin{aligned} k_a &= \frac{1 - \sin 34^\circ}{1 + \sin 34^\circ} \\ &= 0.28 \end{aligned}$$

Calculate the active horizontal earth pressure at the base of the sand layer.

$$\begin{aligned} p_a &= p_v k_a = \gamma H k_a \\ &= \left(108.1\ \frac{\text{lbf}}{\text{ft}^3}\right)(45\ \text{ft})(0.28) \\ &= 1362\ \text{lbf/ft}^2 \end{aligned}$$

The reaction per foot of wall width can be determined.

$$\begin{aligned} R_a &= \tfrac{1}{2} p_a H \\ &= \left(\frac{1}{2}\right)\left(1362\ \frac{\text{lbf}}{\text{ft}^2}\right)(45\ \text{ft}) \\ &= 30{,}645\ \text{lbf/ft of wall width} \quad (30.6\ \text{kips/ft}) \end{aligned}$$

The active resultant for the sand acts at a point one third the distance up from the bottom of the sand or 30 ft below the ground surface.

Perform a static moment analysis about the pinned bottom connection of the sheet pile to determine the tensile force required in the tie rod.

$$\begin{aligned}\sum M_{\text{pin}} &= T_{\text{rod}}L_{\text{rod}} - R_aL \\ &= 0 \\ T_{\text{rod}} &= \frac{\frac{1}{3}R_aH}{L_{\text{rod}}} = \frac{\frac{1}{3}R_aH}{H - 8\text{ ft}} \\ &= \frac{\left(\frac{1}{3}\right)\left(30.6\ \frac{\text{kips}}{\text{ft}}\right)(45\text{ ft})}{45\text{ ft} - 8\text{ ft}} \\ &= 12\text{ kips/ft of wall width}\end{aligned}$$

The answer is (B).

Why Other Options Are Wrong

(A) This solution is incorrect because it neglects to include the installation depth of the deadman anchor in the determination of the tensile force in the rod.

(C) This solution miscalculates the centroid distance.

(D) This solution is incorrect because it assumes the active horizontal pressure distribution for the sand is rectangular shaped instead of triangular.

SOLUTION 2

The Rankine method may be used to find the active earth pressure acting at the base of the wall and granular backfill. Determine the active earth pressure acting at the base of the wall, and then determine the active resultant. The equation for active pressure at the base of the wall is

$$p_a = k_a\sigma_v = k_a\gamma H$$

The equation for the coefficient of active earth pressure is

$$\begin{aligned}k_a &= \frac{1 - \sin\phi}{1 + \sin\phi} = \frac{1 - \sin 31^\circ}{1 + \sin 31^\circ} \\ &= 0.32\end{aligned}$$

Calculate the resultant of the triangular pressure distribution.

$$\begin{aligned}R_a &= \tfrac{1}{2}p_aH = \tfrac{1}{2}k_a\gamma H^2 \\ &= \left(\frac{1}{2}\right)(0.32)\left(118.4\ \frac{\text{lbf}}{\text{ft}^3}\right)(15\text{ ft})^2 \\ &= 4262\text{ lbf/ft of wall length} \quad (4300\text{ lbf/ft})\end{aligned}$$

Note that the resultant acts horizontally at the centroid of the triangular distribution at one third the height of the wall measured upward from the base.

The answer is (C).

Why Other Options Are Wrong

(A) This solution fails to square the height of the wall in the resultant calculation. Note that the units are incorrect as well.

(B) This incorrect solution assumes the active pressure acts at the centroid of the pressure distribution diagram; actually, it is the resultant active force that acts. This answer is obtained by multiplying by a centroid depth of 10 ft rather than 15 ft.

(D) This solution fails to divide the result by two, as required by a triangular distribution.

SOLUTION 3

The Coulomb equation is used for this solution.

$$\begin{aligned}k_a &= \frac{\sin^2(\lambda + \phi)}{\sin^2\lambda\sin(\lambda - \delta)\left(1 + \sqrt{\frac{\sin(\phi + \delta)\sin(\phi - \beta)}{\sin(\lambda - \delta)\sin(\lambda + \beta)}}\right)^2} \\ &= \frac{\sin^2(75^\circ + 31^\circ)}{\begin{array}{l}\sin^2 75^\circ\ \sin(75^\circ - 25^\circ) \\ \quad \times\left(1 + \sqrt{\frac{\sin(31^\circ + 25^\circ)\sin(31^\circ - 15^\circ)}{\sin(75^\circ - 25^\circ)\sin(75^\circ + 15^\circ)}}\right)^2\end{array}} \\ &= 0.54\end{aligned}$$

The active resultant can be calculated.

$$\begin{aligned}R_a &= \tfrac{1}{2}k_a\gamma H^2 \\ &= \left(\frac{1}{2}\right)(0.54)\left(112\ \frac{\text{lbf}}{\text{ft}^3}\right)(22\text{ ft})^2 \\ &= 14{,}636\text{ lbf/ft} \quad (14{,}700\text{ lbf/ft})\end{aligned}$$

The answer is (B).

Why Other Options Are Wrong

(A) This incorrect answer results from using the simplified equation to calculate the coefficient of active earth pressure for a straight wall and non-sloping backfill and then calculating the active resultant.

(C) Confusing the soil-to-structure friction coefficient with the angle of internal friction and reversing them in the calculation results in this incorrect answer.

(D) This solution fails to divide the pressure distribution in half when calculating the active resultant.

SOLUTION 4

The Rankine method may be used to find the active earth pressure acting at the base of the wall and granular backfill. Determine the active earth pressure acting at the base of the wall, and then determine the active resultant. The equation for active pressure at the base of the wall is

$$p_a = k_a\sigma_v = k_a\gamma H$$

The equation for the coefficient of active earth pressure is

$$\begin{aligned} k_a &= \frac{1-\sin\phi}{1+\sin\phi} = \frac{1-\sin 36^\circ}{1+\sin 36^\circ} \\ &= 0.26 \end{aligned}$$

Calculate the resultant of the triangular pressure distribution.

$$\begin{aligned} R_a &= \tfrac{1}{2}p_a H = \tfrac{1}{2}k_a\gamma H^2 \\ &= \left(\frac{1}{2}\right)(0.26)\left(122.6\ \frac{\text{lbf}}{\text{ft}^3}\right)(22\ \text{ft})^2 \\ &= 7714\ \text{lbf/ft of wall length} \quad (7700\ \text{lbf/ft}) \end{aligned}$$

Note that the resultant acts horizontally at the centroid of the triangular distribution at one third the height of the wall measured upward from the base.

The answer is (C).

Why Other Options Are Wrong

(A) This solution neglects to square the height of the wall in the resultant calculation. The units do not work out correctly.

(B) This solution is incorrect because it assumes the active pressure is applied at the centroid of the pressure distribution diagram, where actually, the resultant active force acts. The incorrect solution is obtained by multiplying by the centroid depth of 14.7 ft rather than 22 ft.

(D) This solution fails to divide the result by two, as required by a triangular distribution.

SOLUTION 5

The midpoint depth of the current excavation is 17 ft below the ground surface. Calculate the stability number to determine the appropriate equation to be used for determination of the maximum lateral earth pressure.

$$\begin{aligned} N_o &= \frac{\gamma H}{c} = \frac{\left(101.4\ \frac{\text{lbf}}{\text{ft}^3}\right)(34\ \text{ft})}{400\ \frac{\text{lbf}}{\text{ft}^2}} \\ &= 8.6 \\ N_o &> 6 \end{aligned}$$

The stability number is greater than six; therefore, use the maximum lateral earth pressure equation for soft clay (CH, high-plasticity clay).

$$\begin{aligned} p_{\max} &= \gamma H - 4c \\ &= \left(101.4\ \frac{\text{lbf}}{\text{ft}^3}\right)(34\ \text{ft}) - (4)\left(400\ \frac{\text{lbf}}{\text{ft}^2}\right) \\ &= 1848\ \text{lbf/ft}^2 \quad (1850\ \text{lbf/ft}^2) \end{aligned}$$

The answer is (C).

Why Other Options Are Wrong

(A) This solution mistakenly uses the midpoint of the current excavation depth of 17 ft to determine the expected maximum lateral earth pressure.

(B) This solution mistakenly uses the midpoint of the final excavation depth of 25.5 ft to determine the expected maximum lateral earth pressure.

(D) This solution mistakenly uses the final excavation depth of 51 ft to determine the expected maximum lateral earth pressure.

SOLUTION 6

Use the Rankine method to determine the coefficient of active earth pressure.

$$\begin{aligned} k_a &= \frac{1-\sin\phi}{1+\sin\phi} = \frac{1-\sin 27^\circ}{1+\sin 27^\circ} \\ &= 0.38 \end{aligned}$$

The horizontal active earth pressure at the base of the gravity wall can be calculated for cohesionless soil.

$$\begin{aligned} p_a &= k_a\sigma_v = k_a\gamma_t H \\ &= (0.38)\left(109.1\ \frac{\text{lbf}}{\text{ft}^3}\right)(12\ \text{ft}) \\ &= 497\ \text{lbf/ft}^2 \end{aligned}$$

The horizontal active resultant is located at the centroid of the triangular active pressure distribution along the backfill face. The centroid is located 4 ft above the base of the gravity wall. The active resultant can be calculated as

$$R_{a,h} = \tfrac{1}{2}p_aH = \left(\frac{1}{2}\right)\left(497\ \frac{\text{lbf}}{\text{ft}^2}\right)(12\ \text{ft})$$
$$= 2982\ \text{lbf/ft of wall length}$$

The resistance to sliding can be calculated using the relationship for a keyless foundation base.

$$R_{\text{SL}} = \left(\sum W_i + R_{a,v}\right)\tan\delta + c_AB$$

Due to a vertical backfill face, the vertical resultant does not exist. The resistance to sliding is

$$R_{\text{SL}} = \left(\sum W_i + R_{a,v}\right)\tan\delta + c_AB$$
$$= \left(14{,}000\ \frac{\text{lbf}}{\text{ft}}\right)\tan 20° + \left(150\ \frac{\text{lbf}}{\text{ft}^2}\right)(8\ \text{ft})$$
$$= 6296\ \text{lbf/ft of wall length}$$

The factor of safety against sliding may be determined as

$$F_{\text{SL}} = \frac{R_{\text{SL}}}{R_{a,h}} = \frac{6296\ \frac{\text{lbf}}{\text{ft}}}{2982\ \frac{\text{lbf}}{\text{ft}}}$$
$$= 2.1$$

The answer is (D).

Why Other Options Are Wrong

(A) This incorrect solution is obtained by calculating the horizontal active resultant from a rectangular (constant) pressure distribution rather than from a triangular pressure distribution while using the correct sliding resistance force. The calculated factor of safety is very low compared to the correct factor of safety.

(B) This solution mistakenly uses the friction angle, δ, instead of the angle of internal friction, ϕ, to determine the coefficient of active earth pressure.

(C) This solution neglects to add the adhesion to the resistance-to-sliding equation. The correct factor of safety is greater than this value.

SOLUTION 7

The Rankine method may be used to find the active earth pressures and resultant forces shown. Use the active resultants to find the overturning moment about the toe (point A).

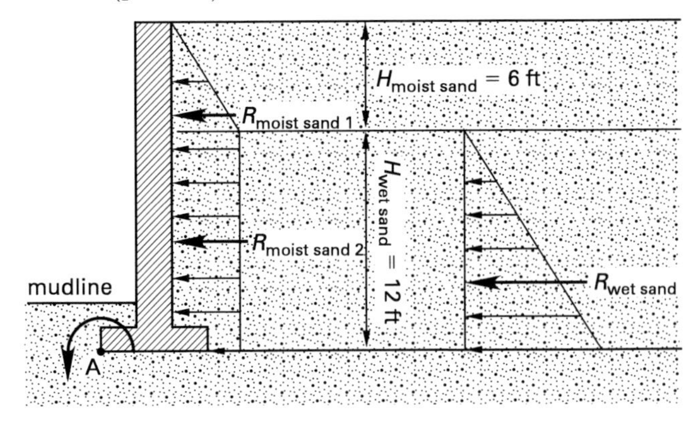

Calculate the coefficient of active earth pressure.

$$k_a = \frac{1-\sin\phi}{1+\sin\phi} = \frac{1-\sin 22°}{1+\sin 22°}$$
$$= 0.45$$

Calculate the horizontal active pressure at the base of the moist sand.

$$\sigma_{h,\text{moist sand}} = k_a\sigma_v$$
$$= k_a\gamma_d(1+w_{\text{moist sand}})H_{\text{moist sand}}$$
$$= (0.45)\left(110.4\ \frac{\text{lbf}}{\text{ft}^3}\right)(1+0.04)(6\ \text{ft})$$
$$= 310\ \text{lbf/ft}^2$$

Calculate the horizontal active resultant for the moist sand.

$$R_{\text{moist sand 1}} = \tfrac{1}{2}\sigma_{h,\text{moist sand}}H_{\text{moist sand}}$$
$$= \left(\frac{1}{2}\right)\left(310\ \frac{\text{lbf}}{\text{ft}^2}\right)(6\ \text{ft})$$
$$= 930\ \text{lbf/ft of wall width}$$

The horizontal active pressure due to the moist sand extends downward into the wet sand layer as a constant distribution. Calculate the horizontal active resultant for the dry sand pressure.

$$R_{\text{moist sand 2}} = \sigma_{h,\text{moist sand}}H_{\text{wet sand}}$$
$$= \left(310\ \frac{\text{lbf}}{\text{ft}^2}\right)(12\ \text{ft})$$
$$= 3720\ \text{lbf/ft of wall}$$

Earth Retaining Structures

Calculate the horizontal active pressure at the base of the wet sand.

$$\begin{aligned}\sigma_{h,\text{wet sand}} &= k_a\sigma_{v,\text{wet sand}} \\ &= k_a\gamma_d(1+w_{\text{wet sand}})H_{\text{wet sand}} \\ &= (0.45)\left(110.4\ \frac{\text{lbf}}{\text{ft}^3}\right)(1+0.15)(12\ \text{ft}) \\ &= 686\ \text{lbf/ft}^2\end{aligned}$$

Calculate the horizontal active resultant for the wet sand.

$$\begin{aligned}R_{\text{wet sand}} &= \tfrac{1}{2}\sigma_{h,\text{wet sand}}H_{\text{wet sand}} \\ &= \left(\frac{1}{2}\right)\left(686\ \frac{\text{lbf}}{\text{ft}^2}\right)(12\ \text{ft}) \\ &= 4116\ \text{lbf/ft of wall width}\end{aligned}$$

Sum the overturning moment about the toe (point A). Use counterclockwise as the positive direction.

$$\begin{aligned}\sum M_{A,\text{OT}} &= \tfrac{1}{3}H_{\text{wet sand}}R_{\text{wet sand}} \\ &\quad + \tfrac{1}{2}H_{\text{wet sand}}R_{\text{moist sand 2}} \\ &\quad + (\tfrac{1}{3}H_{\text{moist sand}} + H_{\text{wet sand}}) \\ &\quad \times R_{\text{moist sand 1}} \\ &= \left(\frac{1}{3}\right)(12\ \text{ft})\left(4116\ \frac{\text{lbf}}{\text{ft}}\right) \\ &\quad + \left(\frac{1}{2}\right)(12\ \text{ft})\left(3720\ \frac{\text{lbf}}{\text{ft}}\right) \\ &\quad + \left(\left(\frac{1}{3}\right)(6\ \text{ft}) + (12\ \text{ft})\right)\left(930\ \frac{\text{lbf}}{\text{ft}}\right) \\ &= 51{,}804\ \text{ft-lbf/ft of wall width}\end{aligned}$$

Calculate the resisting moment about the toe of the retaining wall. The retaining wall self-weight and soil weight above the heel should be determined by breaking the wall and soil into three centroids—a vertical rectangle (wall area 1), a horizontal rectangle (wall base area 2), and another vertical rectangle for the moist sand and wet sand.

$$\begin{aligned}W_{\text{wall area 1}} &= t_{\text{wall}}(H_{\text{moist sand}} + H_{\text{wet sand}} - t_{\text{base}}) \\ &\quad \times \gamma_{\text{concrete}} \\ &= (3\ \text{ft})(6\ \text{ft} + 12\ \text{ft} - 2\ \text{ft})\left(145\ \frac{\text{lbf}}{\text{ft}^3}\right) \\ &= 6960\ \text{lbf/ft of wall length}\end{aligned}$$

$$\begin{aligned}W_{\text{wall base area 2}} &= Bt_{\text{base}}\gamma_{\text{concrete}} \\ &= (6\ \text{ft})(2\ \text{ft})\left(145\ \frac{\text{lbf}}{\text{ft}^3}\right) \\ &= 1740\ \text{lbf/ft of wall length}\end{aligned}$$

$$\begin{aligned}W_{\text{soil area 3}} &= \left(B - \frac{B}{2} - \frac{t_{\text{wall}}}{2}\right)H_{\text{moist sand}}\gamma_d \\ &\quad \times (1 + w_{\text{moist sand}}) \\ &\quad + \left(B - \frac{B}{2} - \frac{t_{\text{wall}}}{2}\right)(H_{\text{wet sand}} - t_{\text{base}})\gamma_d \\ &\quad \times (1 + w_{\text{wet sand}}) \\ &= \left(6\ \text{ft} - \frac{6\ \text{ft}}{2} - \frac{3\ \text{ft}}{2}\right)(6\ \text{ft})\left(110.4\ \frac{\text{lbf}}{\text{ft}^3}\right) \\ &\quad \times (1+0.04) + \left(6\ \text{ft} - \frac{6\ \text{ft}}{2} - \frac{3\ \text{ft}}{2}\right) \\ &\quad \times (12\ \text{ft} - 2\ \text{ft})\left(110.4\ \frac{\text{lbf}}{\text{ft}^3}\right)(1+0.15) \\ &= 2938\ \text{lbf/ft of wall length}\end{aligned}$$

Sum the resisting moment about the toe (point A). Use clockwise as the positive direction.

$$\begin{aligned}\sum M_{A,R} &= \left(\frac{B}{2}\right)W_{\text{wall area 1}} + \left(\frac{B}{2}\right)W_{\text{wall base area 2}} \\ &\quad + \tfrac{7}{8}BW_{\text{soil area 3}} \\ &= \left(\frac{6\ \text{ft}}{2}\right)\left(6960\ \frac{\text{lbf}}{\text{ft}}\right) + \left(\frac{6\ \text{ft}}{2}\right)\left(1740\ \frac{\text{lbf}}{\text{ft}}\right) \\ &\quad + \left(\frac{7}{8}\right)(6\ \text{ft})\left(2938\ \frac{\text{lbf}}{\text{ft}}\right) \\ &= 41{,}525\ \text{ft-lbf/ft}\end{aligned}$$

Calculate the approximate factor of safety against overturning.

$$\begin{aligned}F_{\text{OT}} &= \frac{M_{\text{resisting,toe}}}{M_{\text{overturning,toe}}} = \frac{M_{A,R}}{M_{A,\text{OT}}} \\ &= \frac{41{,}525\ \frac{\text{ft-lbf}}{\text{ft}}}{51{,}804\ \frac{\text{ft-lbf}}{\text{ft}}} \\ &= 0.80\end{aligned}$$

The answer is (B).

Why Other Options Are Wrong

(A) This incorrect solution is obtained by ignoring the wet sand pressure distribution.

Earth Retaining Structures

(C) This incorrect solution is obtained by ignoring the moist sand pressure distribution extending down into the wet sand.

(D) This solution erroneously inverts the resisting moment and the overturning moment in the factor of safety calculation.

SOLUTION 8

The resistance to sliding can be determined from the factor of safety using the required value of 2.

$$F_{\text{SL}} = \frac{R_{\text{SL}}}{R_{a,h}} = 2$$
$$R_{\text{SL}} = F_{\text{SL}} R_{a,h} = 2R_{a,h}$$

Use the Rankine method to determine the coefficient of active earth pressure.

$$k_a = \frac{1-\sin\phi}{1+\sin\phi} = \frac{1-\sin 27°}{1+\sin 27°} = 0.38$$

The horizontal active earth pressure at the base of the gravity wall can be calculated.

$$\begin{aligned} p_a &= k_a\sigma_v = k_a\gamma H_s \\ &= (0.38)\left(105.7\ \frac{\text{lbf}}{\text{ft}^3}\right)(8\ \text{ft}) \\ &= 321.3\ \text{lbf/ft}^2 \end{aligned}$$

The centroid of the triangular active pressure distribution is located at 2.7 ft above the base of the gravity wall. The active resultant can be calculated.

$$\begin{aligned} R_{a,h} &= \tfrac{1}{2}p_a H_s \\ &= \left(\frac{1}{2}\right)\left(321.3\ \frac{\text{lbf}}{\text{ft}^2}\right)(8\ \text{ft}) \\ &= 1285\ \text{lbf/ft of wall length} \end{aligned}$$

Using the equation for the factor of safety, the horizontal active resultant can be used to calculate the wall height through the resistance to sliding equation. The resistance to sliding can be calculated using the relationship for a keyless foundation base.

$$\begin{aligned} R_{\text{SL}} &= \left(\sum W_i + R_{a,v}\right)\tan\delta + c_A B \\ &= 2R_{a,h} \end{aligned}$$

Note that due to a straight backfill face, the vertical resultant does not exist. The weight of the wall can be solved using the resistance to sliding equation and then equated with the density of concrete.

$$\begin{aligned} \sum W_i &= \frac{F_{\text{SL}}R_{a,h} - c_A B}{\tan\delta} \\ &= \gamma_{\text{concrete}} B H_{\text{wall}} \end{aligned}$$

$$\begin{aligned} \gamma_{\text{concrete}} B H_{\text{wall}} &= \frac{F_{\text{SL}}R_{a,h} - c_A B}{\tan\delta} \\ H_{\text{wall}} &= \frac{F_{\text{SL}}R_{a,h} - c_A B}{(\tan\delta)\gamma_{\text{concrete}} B} \\ &= \frac{(2)\left(1285\ \frac{\text{lbf}}{\text{ft}}\right) - \left(150\ \frac{\text{lbf}}{\text{ft}^2}\right)(4\ \text{ft})}{\tan 15°\left(150\ \frac{\text{lbf}}{\text{ft}^3}\right)(4\ \text{ft})} \\ &= 12.3\ \text{ft} \quad (12\ \text{ft}) \end{aligned}$$

The answer is (B).

Why Other Options Are Wrong

(A) This solution mistakenly uses the keyed foundation equation to calculate the resistance to sliding.

(C) This solution mistakenly uses the coefficient of friction, δ, instead of the angle of internal friction, ϕ, to determine the coefficient of active earth pressure.

(D) This solution mistakenly calculates the horizontal active resultant from a rectangular (constant) pressure distribution rather than from a triangular pressure distribution while using the correct sliding resistance force.

SOLUTION 9

The Rankine method is used to find the active earth pressure acting at the base of the soil above the phreatic surface, or water table, and at the base of the wall footing. Determine the active earth pressure acting at the base of the soil above the water table and then determine the active earth pressure (considering buoyancy effects) and hydrostatic pore pressure acting at the base of the footing. Use these values to determine the resultant for each pressure distribution diagram as shown.

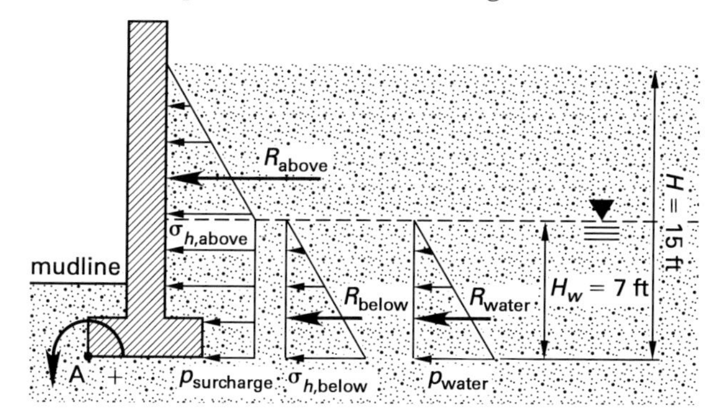

Because the laboratory testing did not reveal significant changes in the strength parameters for the subject soil, the angle of internal friction, ϕ, and coefficient of active earth pressure can be assumed to remain the same.

$$\begin{aligned} k_a &= \frac{1-\sin\phi}{1+\sin\phi} = \frac{1-\sin 31°}{1+\sin 31°} \\ &= 0.32 \end{aligned}$$

The calculation of the horizontal active pressure at the base of the soil layer above the phreatic surface is

$$\begin{aligned} \sigma_{h,\text{above}} &= k_a\sigma_{v,\text{above}} = k_a\gamma(H-H_w) \\ &= (0.32)\left(119.4\ \frac{\text{lbf}}{\text{ft}^3}\right)(15\ \text{ft} - 7\ \text{ft}) \\ &= 306\ \text{lbf/ft}^2 \end{aligned}$$

The horizontal active pressure is treated as a surcharge pressure below the phreatic surface extending as a rectangular pressure distribution to the base of the wall.

$$\sigma_{h,\text{above}} = p_{\text{surcharge}} = 306\ \text{lbf/ft}^2$$

The calculation of the horizontal active pressure (considering buoyancy effects) at the base of the soil layer below the phreatic surface is

$$\begin{aligned} \sigma_{h,\text{below}} &= k_a\sigma_{v,\text{below}} \\ &= k_a\big(\gamma_d(1+w) - \gamma_w\big)H_w \\ &= (0.32)\left(\begin{array}{c}\left(107.6\ \dfrac{\text{lbf}}{\text{ft}^3}\right)(1+0.167) \\ -62.4\ \dfrac{\text{lbf}}{\text{ft}^3}\end{array}\right)(7\ \text{ft}) \\ &= 141.5\ \text{lbf/ft}^2 \end{aligned}$$

The calculation of the hydrostatic pore pressure at the base of the soil layer below the phreatic surface is

$$\begin{aligned} p_{\text{water}} &= \gamma_w H_w \\ &= \left(62.4\ \frac{\text{lbf}}{\text{ft}^3}\right)(7\ \text{ft}) \\ &= 436.8\ \text{lbf/ft}^2 \end{aligned}$$

Because of triangular pressure distributions, each resultant force (per unit length of the retaining wall) may be calculated as

$$\begin{aligned} R_{\text{above}} &= \tfrac{1}{2}\sigma_{h,\text{above}}(H-H_w) \\ &= \left(\frac{1}{2}\right)\left(306\ \frac{\text{lbf}}{\text{ft}^2}\right)(15\ \text{ft} - 7\ \text{ft}) \\ &= 1224\ \text{lbf/ft of wall length} \end{aligned}$$

$$\begin{aligned} R_{\text{surcharge}} &= p_{\text{surcharge}}H_w \\ &= \left(306\ \frac{\text{lbf}}{\text{ft}^2}\right)(7\ \text{ft}) \\ &= 2142\ \text{lbf/ft of wall length} \end{aligned}$$

$$\begin{aligned} R_{\text{below}} &= \tfrac{1}{2}\sigma_{h,\text{below}}H_w \\ &= \left(\frac{1}{2}\right)\left(141.5\ \frac{\text{lbf}}{\text{ft}^2}\right)(7\ \text{ft}) \\ &= 495\ \text{lbf/ft of wall length} \end{aligned}$$

$$\begin{aligned} R_{\text{water}} &= \tfrac{1}{2}p_{\text{water}}H_w \\ &= \left(\frac{1}{2}\right)\left(436.8\ \frac{\text{lbf}}{\text{ft}^2}\right)(7\ \text{ft}) \\ &= 1529\ \text{lbf/ft of wall length} \end{aligned}$$

The resultant values for the soil (including buoyant forces) below the phreatic surface and the hydrostatic conditions, respectively, should be summed.

$$\begin{aligned} R_{\text{below,total}} &= R_{\text{below}} + R_{\text{water}} \\ &= 495\ \frac{\text{lbf}}{\text{ft}} + 1529\ \frac{\text{lbf}}{\text{ft}} \\ &= 2024\ \text{lbf/ft of wall length} \end{aligned}$$

The resultant forces act at one third of the height up from the base of their respective pressure distribution diagrams. The perpendicular distances from the axis of rotation (lever arms) must be used to determine the total overturning moment while keeping in mind that the calculations are per unit length of wall. Use counterclockwise as the positive direction.

$$\begin{aligned} M_A &= \left(H_w + \tfrac{1}{3}(H - H_w)\right)R_{\text{above}} \\ &\quad + \tfrac{1}{2}H_w R_{\text{surcharge}} + \tfrac{1}{3}H_w R_{\text{below,total}} \\ &= \left(7\text{ ft} + \left(\frac{1}{3}\right)(15\text{ ft} - 7\text{ ft})\right)\left(1224\ \frac{\text{lbf}}{\text{ft}}\right) \\ &\quad + \left(\frac{1}{2}\right)(7\text{ ft})\left(2142\ \frac{\text{lbf}}{\text{ft}}\right) \\ &\quad + \left(\frac{1}{3}\right)(7\text{ ft})\left(2024\ \frac{\text{lbf}}{\text{ft}}\right) \\ &= 24{,}000\text{ ft-lbf/ft of wall length} \quad (24\text{ ft-kips/ft}) \end{aligned}$$

The answer is (B).

Why Other Options Are Wrong

(A) This solution erroneously skips the step that calculates the active pressure. Instead, the total unit weight is used to determine the active resultant force for each case, and the incorrect equation, $\frac{1}{2}\sigma_a H$, is used instead of $\frac{1}{2}k_a\gamma H^2$.

(C) This incorrect solution adjusts the horizontal active pressure for buoyancy effects and then neglects to include the horizontal pressure due to water.

(D) This solution uses the wrong centroid distances.

SOLUTION 10

The Rankine method may be used to find the active earth pressure resultant because of a straight back with granular/cohesionless backfill. Calculate the active earth pressure acting at the base of the wall due to the backfill.

$$p_a = k_a\sigma_v = k_a\gamma H$$

To determine the resulting uniform pressure acting on the wall due to the surcharge pressure in addition to the backfill, use the formula

$$p_q = k_a q$$

Use the equation for the coefficient of active earth pressure.

$$\begin{aligned} k_a &= \frac{1 - \sin\phi}{1 + \sin\phi} \\ &= \frac{1 - \sin 35^\circ}{1 + \sin 35^\circ} \\ &= 0.27 \end{aligned}$$

Calculate the resultant of the triangular pressure distribution.

$$\begin{aligned} R_a &= \tfrac{1}{2}p_a H = \tfrac{1}{2}k_a\gamma H^2 \\ &= \left(\frac{1}{2}\right)(0.27)\left(123.1\ \frac{\text{lbf}}{\text{ft}^3}\right)(18\text{ ft})^2 \\ &= 5384\text{ lbf/ft of wall length} \end{aligned}$$

The resultant acts horizontally at one third the distance up the height of the wall from the base.

Calculate the resultant of the uniform pressure distribution due to the surcharge.

$$\begin{aligned} R_q &= p_q H = k_a q H \\ &= (0.27)\left(148\ \frac{\text{lbf}}{\text{ft}^2}\right)(18\text{ ft}) \\ &= 719\text{ lbf/ft of wall length} \end{aligned}$$

The resultant acts horizontally at half the distance up the height of the wall from the base.

The overturning moment is calculated about the toe using the perpendicular distances from the resultant forces to the axis of rotation (lever arms), keeping in mind that the calculations are per unit length of wall. For simplicity, use counterclockwise as the positive direction.

$$\begin{aligned} M_{\text{toe}} &= \tfrac{1}{3}HR_a + \tfrac{1}{2}HR_q \\ &= \left(\frac{1}{3}\right)(18\text{ ft})(5384\text{ lbf}) + \left(\frac{1}{2}\right)(18\text{ ft})(719\text{ lbf}) \\ &= 38{,}800\text{ ft-lbf} \quad (39\text{ ft-kips}) \end{aligned}$$

The answer is (C).

Why Other Options Are Wrong

(A) This solution fails to square the height of the wall in the resultant calculation (the units are incorrect as well) and proceeds to determine the overturning moment.

(B) A faulty assumption that the active pressure acts at the centroid of the pressure distribution diagram (where actually, the resultant active force acts) produces this incorrect answer. The incorrect solution is obtained by multiplying by the centroid depth of 12 ft rather than 18 ft.

(D) Failure to convert the surcharge pressure to a uniform pressure acting on the retaining wall backfill face and leaving it as the given surcharge pressure value results in this incorrect answer.

SOLUTION 11

Use the Rankine method to determine the coefficient of active earth pressure.

$$\begin{aligned} k_a &= \frac{1-\sin\phi}{1+\sin\phi} = \frac{1-\sin 32°}{1+\sin 32°} \\ &= 0.31 \end{aligned}$$

The horizontal active earth pressure at the base of the gravity wall can be calculated.

$$\begin{aligned} p_a &= k_a\sigma_v = k_a\gamma_t H \\ &= (0.31)\left(124.4\ \frac{\text{lbf}}{\text{ft}^3}\right)(18\ \text{ft}) \\ &= 694\ \text{lbf/ft}^2 \end{aligned}$$

The horizontal active resultant is located at the centroid of the triangular active pressure distribution along the backfill face. The centroid is located 6 ft above the base of the gravity wall. The active resultant can be calculated.

$$\begin{aligned} R_{a,h} &= \tfrac{1}{2}p_a H \\ &= \left(\frac{1}{2}\right)\left(694\ \frac{\text{lbf}}{\text{ft}^2}\right)(18\ \text{ft}) \\ &= 6246\ \text{lbf/ft of wall length} \end{aligned}$$

The resistance to sliding can be calculated using the relationship for a keyed foundation base.

$$R_{\text{SL}} = \left(\sum W_i + R_{a,v}\right)\tan\phi + c_A B$$

Note that due to a straight backfill face, the vertical resultant does not exist. The resistance to sliding can be calculated.

$$\begin{aligned} R_{\text{SL}} &= \left(\sum W_i + R_{a,v}\right)\tan\phi + c_A B \\ &= \left(16{,}500\ \frac{\text{lbf}}{\text{ft}}\right)\tan 32° + \left(250\ \frac{\text{lbf}}{\text{ft}^2}\right)(12\ \text{ft}) \\ &= 13{,}310\ \text{lbf/ft of wall length}\quad (13\ \text{kips/ft}) \end{aligned}$$

The answer is (D).

Why Other Options Are Wrong

(A) This incorrect solution uses the keyless foundation equation to calculate the resistance to sliding.

(B) This solution mistakenly uses the friction angle, δ, instead of the angle of internal friction, ϕ, to determine the coefficient of active earth pressure.

(C) This incorrect solution calculates the horizontal active resultant from a rectangular (constant) pressure distribution rather than from a triangular pressure distribution while using the correct sliding resistance force.

SOLUTION 12

The Rankine method may be used to find the active earth pressure acting at the base of the wall footing. Use the active earth pressure value to determine the active resultant and overturning moment about the toe (point A).

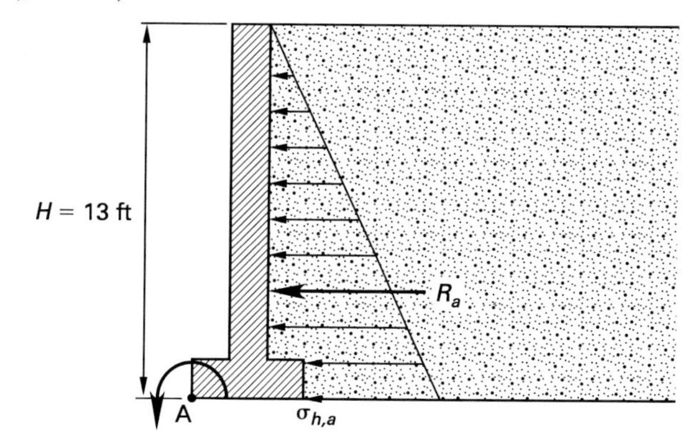

Calculate the coefficient of active earth pressure.

$$\begin{aligned} k_a &= \frac{1-\sin\phi}{1+\sin\phi} = \frac{1-\sin 33°}{1+\sin 33°} \\ &= 0.29 \end{aligned}$$

Calculate the horizontal active pressure at the base of the wall.

$$\begin{aligned} \sigma_h &= k_a\sigma_v = k_a\gamma H \\ &= (0.29)\left(113.8\ \frac{\text{lbf}}{\text{ft}^3}\right)(13\ \text{ft}) \\ &= 429\ \text{lbf/ft}^2 \end{aligned}$$

The horizontal active pressure is the bottom value of the triangular pressure distribution at the base of the wall. Determine the active resultant.

$$\begin{aligned} R_a &= \frac{1}{2}\sigma_h H \\ &= \left(\frac{1}{2}\right)\left(429\ \frac{\text{lbf}}{\text{ft}^2}\right)(13\ \text{ft}) \\ &= 2789\ \text{lbf/ft of wall width} \end{aligned}$$

The active resultant force acts one third of the height above the base of the pressure distribution diagram. The perpendicular distance from the axis of rotation (lever arm) must be used to determine the total overturning moment. The calculations are determined per unit length of wall. Use counterclockwise as the positive direction.

$$\begin{aligned} M_{A,\mathrm{OT}} &= \tfrac{1}{3}HR_a \\ &= \left(\frac{1}{3}\right)(13\ \mathrm{ft})\left(2789\ \frac{\mathrm{lbf}}{\mathrm{ft\ of\ wall\ width}}\right) \\ &= 12{,}086\ \mathrm{ft\text{-}lbf\ per\ ft\ of\ wall\ width}\quad (12\ \mathrm{ft\text{-}kips}) \end{aligned}$$

The answer is (B).

Why Other Options Are Wrong

(A) This solution erroneously skips the step that calculates the active pressure. Instead, the total unit weight is used to determine the active resultant force. The units do not work out correctly.

(C) This incorrect solution assumes the active resultant acts at the centroid of a rectangular distribution rather than at the centroid of a triangular distribution.

(D) This incorrect solution assumes the active resultant acts at the top of the triangular distribution rather than at one third the height of the triangular distribution.

SOLUTION 13

Coulomb theory should be used to find the active earth pressure acting at the base of the wall and granular backfill. Determine the vertical and horizontal active earth pressure components acting at the centroid of the pressure distribution on the wall. The equation for vertical and horizontal active earth pressure components can be obtained from published sources.

$$R_{a,v} = 0.5k_vH^2$$

$$R_{a,h} = 0.5k_hH^2$$

The vertical and horizontal active earth pressure components can also be found from published sources; $k_v = 10\ \mathrm{lbf/ft^2}$ and $k_h = 45\ \mathrm{lbf/ft^2}$ per foot of wall length.

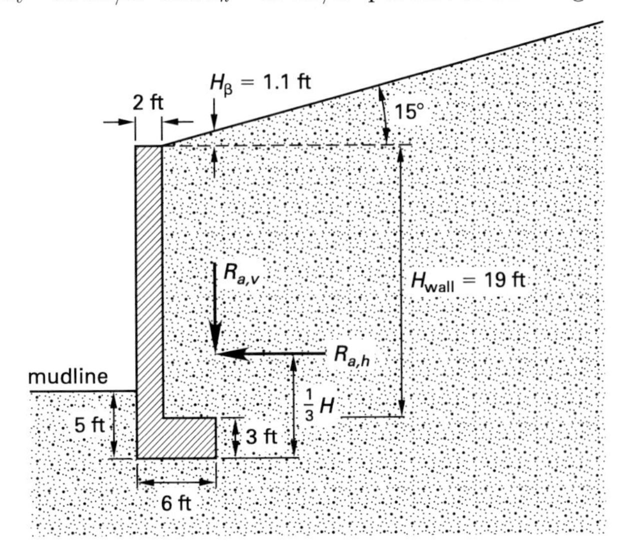

Calculate the active resultant forces at the base of the wall.

$$\begin{aligned} R_{a,v} &= 0.5k_v(t_{\mathrm{base}} + H_{\mathrm{wall}} + H_\beta)^2 \\ &= (0.5)\left(10\ \frac{\mathrm{lbf}}{\mathrm{ft}^2}\right)(3\ \mathrm{ft} + 19\ \mathrm{ft} + (4\ \mathrm{ft})\tan 15°)^2 \\ &= 2661.5\ \mathrm{lbf/ft\ of\ wall\ length}\quad (2.7\ \mathrm{kips/ft}) \end{aligned}$$

$$\begin{aligned} R_{a,h} &= 0.5k_h(t_{\mathrm{base}} + H_{\mathrm{wall}} + H_\beta)^2 \\ &= (0.5)\left(45\ \frac{\mathrm{lbf}}{\mathrm{ft}^2}\right)(3\ \mathrm{ft} + 19\ \mathrm{ft} + (4\ \mathrm{ft})\tan 15°)^2 \\ &= 11{,}976.9\ \mathrm{lbf/ft\ of\ wall\ length}\quad (12\ \mathrm{kips/ft}) \end{aligned}$$

The horizontal resultant acts at the centroid of the triangular distribution at one third the height of the wall measured upward from the base. Calculate the overturning moment about the toe of the retaining wall.

$$\begin{aligned} M_{\mathrm{overturning,toe}} &= \tfrac{1}{3}HR_{a,h} \\ &= \left(\frac{1}{3}\right)(23.1\ \mathrm{ft})\left(12\ \frac{\mathrm{kips}}{\mathrm{ft}}\right) \\ &= 92.3\ \mathrm{ft\text{-}kips/ft\ of\ wall\ length} \end{aligned}$$

Calculate the resisting moment about the toe of the retaining wall. The resisting moment consists of the retaining wall weight and the soil weight. The retaining wall weight should be determined by breaking the structure into two parts—a vertical rectangle for the weight of the wall and a horizontal rectangle for the weight of the base.

$$\begin{aligned} W_{\text{wall}} &= t_{\text{wall}} H_{\text{wall}} \gamma_{\text{concrete}} \\ &= (2 \text{ ft})(19 \text{ ft})\left(145 \ \frac{\text{lbf}}{\text{ft}^3}\right) \\ &= 5510 \text{ lbf/ft of wall length} \quad (5.5 \text{ kips/ft}) \end{aligned}$$

$$\begin{aligned} W_{\text{base}} &= t_{\text{base}} B \gamma_{\text{concrete}} \\ &= (3 \text{ ft})(6 \text{ ft})\left(145 \ \frac{\text{lbf}}{\text{ft}^3}\right) \\ &= 2610 \text{ lbf/ft of wall length} \quad (2.6 \text{ kips/ft}) \end{aligned}$$

The weight of backfill soil above the heel can be calculated as a rectangle (soil area 1) and a small triangle in the sloped backfill above the top of the wall (soil area 2).

$$\begin{aligned} W_{\text{soil area 1}} &= (B - t_{\text{wall}}) H_{\text{wall}} \gamma_d (1 + w) \\ &= (6 \text{ ft} - 2 \text{ ft})(19 \text{ ft})\left(119.1 \ \frac{\text{lbf}}{\text{ft}^3}\right) \\ &\quad \times (1 + 0.133) \\ &= 10{,}255 \text{ lbf/ft of wall length} \\ &\quad (10.3 \text{ kips/ft}) \end{aligned}$$

$$\begin{aligned} W_{\text{soil area 2}} &= 0.5(B - t_{\text{wall}}) H_\beta \gamma_d (1 + w) \\ &= (0.5)(6 \text{ ft} - 2 \text{ ft})\big((4 \text{ ft})\tan 15°\big) \\ &\quad \times \left(119.1 \ \frac{\text{lbf}}{\text{ft}^3}\right)(1 + 0.133) \\ &= 289 \text{ lbf/ft of wall length} \quad (0.3 \text{ kips/ft}) \end{aligned}$$

$$\begin{aligned} M_{\text{resisting,toe}} &= 0.5 t_{\text{wall}} W_{\text{wall}} + (0.5(B - t_{\text{wall}}) + t_{\text{wall}}) \\ &\quad \times (W_{\text{soil area 1}}) + \tfrac{1}{2} B W_{\text{base}} \\ &\quad + \left(\frac{2}{3}(B - t_{\text{wall}}) + t_{\text{wall}}\right) W_{\text{soil area 2}} \\ &\quad + B R_{a,v} \\ &= (0.5)(2 \text{ ft})\left(5.5 \ \frac{\text{kips}}{\text{ft}}\right) \\ &\quad + \big((0.5)(6 \text{ ft} - 2 \text{ ft}) + 2 \text{ ft}\big) \\ &\quad \times \left(10.3 \ \frac{\text{kips}}{\text{ft}}\right) + \left(\frac{1}{2}\right)(6 \text{ ft}) \\ &\quad \times \left(2.6 \ \frac{\text{kips}}{\text{ft}}\right) \\ &\quad + \left(\left(\frac{2}{3}\right)(6 \text{ ft} - 2 \text{ ft}) + 2 \text{ ft}\right) \\ &\quad \times \left(0.3 \ \frac{\text{kips}}{\text{ft}}\right) + (6 \text{ ft})\left(2.7 \ \frac{\text{kips}}{\text{ft}}\right) \\ &= 72.1 \text{ ft-kips/ft of wall length} \end{aligned}$$

$$\begin{aligned} F_{\text{OT}} &= \frac{M_{\text{resisting,toe}}}{M_{\text{overturning,toe}}} = \frac{72.1 \ \frac{\text{ft-kips}}{\text{ft}}}{92.3 \ \frac{\text{ft-kips}}{\text{ft}}} \\ &= 0.78 \end{aligned}$$

The answer is (B).

Why Other Options Are Wrong

(A) This solution is incorrect because it uses Rankine theory instead of Coulomb theory, neglecting the vertical active component and using a triangular distribution against the retaining wall backfill face.

(C) This solution mistakenly inverts the factor of safety equation.

(D) This solution is incorrect because it interchanges the vertical and horizontal active components used in calculating the resultant values.

SOLUTION 14

The clay layer is normally consolidated, and the thickness has been confirmed at 6 ft. The equation for consolidation settlement can be used to determine any settlement that results from construction of the retaining wall and backfill.

$$S = C_c \left(\frac{H_o}{1 + e_o}\right) \log_{10} \frac{\sigma'_{vo} + \Delta\sigma_v}{\sigma'_{vo}}$$

Earth Retaining Structures

Determine the initial effective stress at the midpoint of the clay layer by determining the initial total stress halfway through the clay layer and subtracting the pore pressure at that point.

$$\sigma'_{vo} = \sigma_{\text{sand,dry}} + \sigma_{\text{sand,wet}} + \sigma_{\text{clay(midpoint)}} - u$$

$$\begin{aligned}\sigma_{\text{sand,dry}} &= \gamma_d D_1 = \left(106.5\ \frac{\text{lbf}}{\text{ft}^3}\right)(22\ \text{ft})\\ &= 2343\ \text{lbf/ft}^2\end{aligned}$$

$$\begin{aligned}\sigma_{\text{sand,wet}} &= \gamma_d w D_1 = \left(106.5\ \frac{\text{lbf}}{\text{ft}^3}\right)(1+0.173)(6\ \text{ft})\\ &= 750\ \text{lbf/ft}^2\end{aligned}$$

$$\begin{aligned}\sigma_{\text{clay(midpoint)}} &= \gamma_d\left(\frac{D_2}{2}\right) = \left(127.8\ \frac{\text{lbf}}{\text{ft}^3}\right)\left(\frac{6\ \text{ft}}{2}\right)\\ &= 383\ \text{lbf/ft}^2\end{aligned}$$

$$u = \gamma_w H = \left(62.4\ \frac{\text{lbf}}{\text{ft}^3}\right)(9\ \text{ft}) = 562\ \text{lbf/ft}^2$$

$$\begin{aligned}\sigma'_{vo} &= \sigma_{\text{sand,dry}} + \sigma_{\text{sand,wet}} + \sigma_{\text{clay(midpoint)}} - u\\ &= 2343\ \frac{\text{lbf}}{\text{ft}^2} + 750\ \frac{\text{lbf}}{\text{ft}^2} + 383\ \frac{\text{lbf}}{\text{ft}^2} - 562\ \frac{\text{lbf}}{\text{ft}^2}\\ &= 2914\ \text{lbf/ft}^2\end{aligned}$$

Determine the new effective stress. To evaluate the correct conditions, the new retaining wall, backfill, and surcharge should replace the soil profile properties given in the problem statement to a depth equivalent to the bearing depth of the retaining wall footing.

$$\Delta\sigma_v = p_q + \sigma_{\text{backfill(18 ft)}} - \sigma_{\text{sand,dry(18 ft)}}$$

$$\begin{aligned}\sigma_{\text{backfill(18 ft)}} &= \gamma_{\text{BF}} H_{\text{BF}} = \left(118.5\ \frac{\text{lbf}}{\text{ft}^3}\right)(18\ \text{ft})\\ &= 2133\ \text{lbf/ft}^2\end{aligned}$$

$$\begin{aligned}\sigma_{\text{sand,dry(18 ft)}} &= \gamma_D H_{\text{BF}} = \left(106.5\ \frac{\text{lbf}}{\text{ft}^3}\right)(18\ \text{ft})\\ &= 1917\ \text{lbf/ft}^2\end{aligned}$$

$$\begin{aligned}\Delta\sigma_v &= p_q + \sigma_{\text{backfill(18 ft)}} - \sigma_{\text{sand,dry(18 ft)}}\\ &= 148\ \frac{\text{lbf}}{\text{ft}^2} + 2133\ \frac{\text{lbf}}{\text{ft}^2} - 1917\ \frac{\text{lbf}}{\text{ft}^2}\\ &= 364\ \text{lbf/ft}^2\end{aligned}$$

Determine the total expected settlement due to placement of the surcharge.

$$\begin{aligned}S &= C_c\left(\frac{H_o}{1+e_o}\right)\log_{10}\frac{\sigma'_{vo}+\Delta\sigma_v}{\sigma'_{vo}} = (0.570)\left(\frac{6\ \text{ft}}{1+0.300}\right)\\ &\quad\times\log_{10}\left(\frac{2914\ \frac{\text{lbf}}{\text{ft}^2} + 364\ \frac{\text{lbf}}{\text{ft}^2}}{2914\ \frac{\text{lbf}}{\text{ft}^2}}\right)\left(12\ \frac{\text{in}}{\text{ft}}\right)\\ &= 1.6\ \text{in}\quad (2\ \text{in})\end{aligned}$$

The answer is (B).

Why Other Options Are Wrong

(A) This solution mistakenly uses the given coefficient of vertical consolidation value in the settlement equation instead of the coefficient of compression value. This result has incorrect units and shows there is no settlement.

(C) This incorrect answer is obtained if the initial void ratio is subtracted from one rather than added to one—a common error. This result is very high compared to the correct answer.

(D) A common error is to divide by the initial void ratio rather than to first add the initial void ratio to one. This incorrect result is very high compared to the correct answer.

SOLUTION 15

The tension development along each strip can be determined using the coefficient of active earth pressure. The tension value is based on a contributory area equivalent to the lateral and vertical spacing.

$$T_i = \gamma d_i h s k_a$$

The coefficient of active earth pressure can be calculated.

$$\begin{aligned}k_a &= \frac{1-\sin\phi}{1+\sin\phi} = \frac{1-\sin 32^\circ}{1+\sin 32^\circ}\\ &= 0.31\end{aligned}$$

The horizontal spacing is 2 ft on center and the vertical height interval is 2 ft. The tension in the first strip, T_1, can be determined. The remaining tension values can be tabulated.

$$\begin{aligned}T_1 &= \gamma d_1 h s k_a\\ &= \left(119.3\ \frac{\text{lbf}}{\text{ft}^3}\right)(1\ \text{ft})(2\ \text{ft})(2\ \text{ft})(0.31)\\ &= 148\ \text{lbf}\end{aligned}$$

strip	depth to strip (ft)	tension (lbf)
1	1	148
2	3	444
3	5	740
4	7	1036
5	9	1331
6	11	1627
7	13	1923
8	15	2219
9	17	2515
10	19	2811
11	21	3107
12	23	3402
	$\sum T_i$ (lbf)	21,303

The Rankine method may be used to find the active earth pressure resultant force acting on the wall from the granular backfill. Determine the active earth pressure acting at the base of the wall and then determine the active resultant (per foot of wall width).

$$p_a = k_a \sigma_v = k_a \gamma H$$

Calculate the resultant of the triangular pressure distribution.

$$\begin{aligned} R_{a,h} &= \tfrac{1}{2} p_a H = \tfrac{1}{2} k_a \gamma H^2 \\ &= \left(\frac{1}{2}\right)(0.31)\left(119.3\ \frac{\text{lbf}}{\text{ft}^3}\right)(24\ \text{ft})^2 \\ &= 10{,}651\ \text{lbf/ft of wall width} \end{aligned}$$

The sum of the tension force from each strip should be equivalent to the horizontal component of the active earth pressure resultant force, with consideration for the lateral 2 ft strip spacing.

$$\begin{aligned} \frac{\sum T_i}{s} &= \frac{21{,}303\ \text{lbf}}{2\ \text{ft}} = 10{,}652\ \text{lbf/ft} \\ &= R_{a,h} \end{aligned}$$

The length dimension of each reinforcing strip is a sum of the effective length (beyond the Rankine active zone) and the length within the Rankine active zone.

$$L_T = L_e + L_R$$

Since both sides of the reinforcing strip contribute to the development of the frictional resistance, calculate the effective length using twice the width value. The minimum effective length should occur at the uppermost strip. Select the tension and depth combination for the uppermost strip, and check other value combinations to confirm that the effective length is consistent.

$$\begin{aligned} T_i &= 2bL_e f \gamma d_i = 2bL_e(\tan\delta)\gamma d_i \\ L_e &= \frac{T_i}{2b(\tan\delta)\gamma d_i} \\ &= \frac{T_i}{(2)\left((3\ \text{in})\left(\frac{1\ \text{ft}}{12\ \text{in}}\right)\right)\tan 20^\circ\left(119.3\ \frac{\text{lbf}}{\text{ft}^3}\right)d_i} \\ &= \frac{T_i}{\left(21.8\ \frac{\text{lbf}}{\text{ft}^2}\right)d_i} \end{aligned}$$

$$\begin{aligned} L_{e,1} &= \frac{T_1}{\left(21.8\ \frac{\text{lbf}}{\text{ft}^2}\right)d_1} = \frac{148\ \text{lbf}}{\left(21.8\ \frac{\text{lbf}}{\text{ft}^2}\right)(1\ \text{ft})} \\ &= 6.8\ \text{ft} \end{aligned}$$

$$\begin{aligned} L_{e,12} &= \frac{T_{12}}{\left(21.8\ \frac{\text{lbf}}{\text{ft}^2}\right)d_{12}} = \frac{3402\ \text{lbf}}{\left(21.8\ \frac{\text{lbf}}{\text{ft}^2}\right)(23\ \text{ft})} \\ &= 6.8\ \text{ft} \end{aligned}$$

To determine the length within the Rankine active zone, the angle of the Rankine failure plane with horizontal should be determined. The width of the Rankine active zone at the widest point (near the top of the backfill) can be calculated. This value is made equivalent to the length of the strip in the Rankine active zone, L_R.

$$\begin{aligned} \theta &= 45^\circ + \frac{\phi}{2} = 45^\circ + \frac{32^\circ}{2} \\ &= 61^\circ \\ L_R &= H\tan(90^\circ - \theta) \\ &= (24\ \text{ft})\tan(90^\circ - 61^\circ) \\ &= 13.3\ \text{ft} \end{aligned}$$

The total reinforcing strip length can be calculated.

$$\begin{aligned} L_T &= L_e + L_R = 6.8\ \text{ft} + 13.3\ \text{ft} \\ &= 20.1\ \text{ft} \quad (20\ \text{ft}) \end{aligned}$$

The answer is (C).

Why Other Options Are Wrong

(A) This solution neglects to include the length of the strip passing through the Rankine active zone.

(B) This solution mistakenly uses the angle of internal friction for the backfill instead of the material friction angle.

(D) This solution mistakenly uses only one side of the reinforcing strip to calculate the development of the frictional resistance.

SOLUTION 16

The tension development in rod no. 8 can be determined using the coefficient of active earth pressure. The tension value is per foot of wall width and is based on a contributory area equivalent to the lateral and vertical spacing.

$$T_i = \gamma d_i hsk_a = \big(\gamma_d(1+w)\big)d_8 hsk_a$$

The coefficient of active earth pressure can be calculated as

$$k_a = \frac{1-\sin\phi}{1+\sin\phi} = \frac{1-\sin 28^\circ}{1+\sin 28^\circ} = 0.36$$

The horizontal (lateral) spacing is 3 ft on center and the vertical interval is 2 ft on center. The tension in the eighth rod, T_8, can be determined.

$$\begin{aligned} T_8 &= \big(\gamma_d(1+w)\big)d_8 hsk_a \\ &= \left(104.7\ \frac{\text{lbf}}{\text{ft}^3}\right)(1+0.17)(15\ \text{ft})(3\ \text{ft})(2\ \text{ft})(0.36) \\ &= 3969\ \text{lbf}\quad (4.0\ \text{kips}) \end{aligned}$$

The answer is (C).

Why Other Options Are Wrong

(A) This solution mistakenly calculates the tension in the first rod.

(B) This solution mistakenly includes the friction angle in calculating the required tension of rod no. 8.

(D) This solution mistakenly uses the coefficient of passive earth pressure rather than the active pressure coefficient in the calculation of the tension in the eighth rod.

SOLUTION 17

The active horizontal earth pressure can be calculated at the cut line elevation. The Rankine equation for the active earth pressure coefficient may be used.

$$\begin{aligned} p_{a,\text{cut}} &= p_v k_a = \gamma(1+w)H\left(\frac{1-\sin\phi}{1+\sin\phi}\right) \\ &= \left(114.1\ \frac{\text{lbf}}{\text{ft}^3}\right)(1+0.06)(20\ \text{ft})\left(\frac{1-\sin 29^\circ}{1+\sin 29^\circ}\right) \\ &= 839\ \text{lbf/ft}^2 \end{aligned}$$

The net passive horizontal earth pressure at the base of the bulkhead can be calculated as the difference between the total passive pressure and the total active pressure.

$$\begin{aligned} p_{p,\text{bulkhead base}} &= k_p\gamma_d(1+w)D - k_a\gamma_d(1+w)(H+D) \\ &= \left(\frac{1+\sin\phi}{1-\sin\phi}\right)\gamma_d(1+w)D \\ &\quad -\left(\frac{1-\sin\phi}{1+\sin\phi}\right)\gamma_d(1+w)(H+D) \\ &= \left(\frac{1+\sin 29^\circ}{1-\sin 29^\circ}\right)\left(114.1\ \frac{\text{lbf}}{\text{ft}^3}\right) \\ &\quad \times(1+0.06)(15\ \text{ft}) \\ &\quad -\left(\frac{1-\sin 29^\circ}{1+\sin 29^\circ}\right)\left(114.1\ \frac{\text{lbf}}{\text{ft}^3}\right) \\ &\quad \times(1+0.06)(35\ \text{ft}) \\ &= 3760\ \text{lbf/ft}^2 \end{aligned}$$

Using static equilibrium, the pivot point can be calculated by comparing areas. The height, h, of the pivot point above the bottom tip of the bulkhead can be determined.

$$\tfrac{1}{2}p_{a,\text{cut}}(D-h) = \tfrac{1}{2}p_{p,\text{bulkhead base}}h$$

$$\left(\frac{1}{2}\right)\left(839\ \frac{\text{lbf}}{\text{ft}^2}\right)(15\ \text{ft}-h) = \left(\frac{1}{2}\right)\left(3760\ \frac{\text{lbf}}{\text{ft}^2}\right)h$$

Solving for h,

$$\begin{aligned} h &= \frac{\left(839\ \frac{\text{lbf}}{\text{ft}^2}\right)(15\ \text{ft})}{4599\ \frac{\text{lbf}}{\text{ft}^2}} \\ &= 2.7\ \text{ft} \end{aligned}$$

Ignoring the passive pressure, perform a static moment analysis, clockwise positive, about the pivot point to determine the tensile force required in the tie rod.

$$\begin{aligned} \sum M_{\text{pivot point}} &= T_{\text{rod}}L_{\text{rod}} - R_{a,\text{sand}}L_{\text{sand}} \\ &= 0 \end{aligned}$$

Earth Retaining Structures

$$
\begin{aligned}
T_{\text{rod}} &= \frac{R_{a,\text{sand}}\left(\frac{1}{3}H + D - h\right)}{H + D - 5\ \text{ft} - h} \\
&= \frac{\frac{1}{2}p_{a,\text{cut}}H\left(\frac{1}{3}H + D - h\right)}{H + D - 5\ \text{ft} - 2.7\ \text{ft}} \\
&= \frac{\left(\frac{1}{2}\right)\left(839\ \frac{\text{lbf}}{\text{ft}^2}\right)(20\ \text{ft}) \times \left(\left(\frac{1}{3}\right)(20\ \text{ft}) + 15\ \text{ft} - 2.7\ \text{ft}\right)}{30\ \text{ft} - 2.7\ \text{ft}} \\
&= 5829\ \text{lbf/ft of bulkhead width} \quad (5.8\ \text{kips/ft})
\end{aligned}
$$

The answer is (B).

Why Other Options Are Wrong

(A) This solution neglects to include the installation depth of the deadman anchor in the determination of the tensile force in the rod.

(C) This solution is incorrect because it sums the moment about the bottom tip of the bulkhead rather than about the pivot point.

(D) This solution neglects to calculate the distance from the pivot point to the centroid of the active pressure distribution.

9 Shallow Foundations

BEARING CAPACITY

PROBLEM 1

Using the Terzaghi-Meyerhof bearing capacity equation and Terzaghi bearing capacity factors, what is most nearly the ultimate bearing capacity for the continuous footing shown?

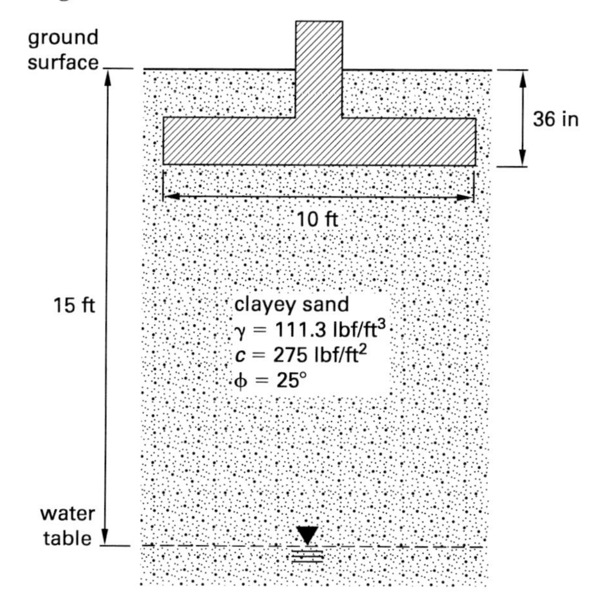

(A) $8300\ \text{lbf/ft}^2$

(B) $13{,}500\ \text{lbf/ft}^2$

(C) $16{,}500\ \text{lbf/ft}^2$

(D) $21{,}900\ \text{lbf/ft}^2$

Hint: The zone of influence due to the footing load extends to a depth equivalent to approximately the footing width.

PROBLEM 2

What is most nearly the allowable bearing pressure of the given shallow circular footing in sand? Use a factor of safety equal to 2 and the Terzaghi-Meyerhof equation, and ignore corrections for overburden.

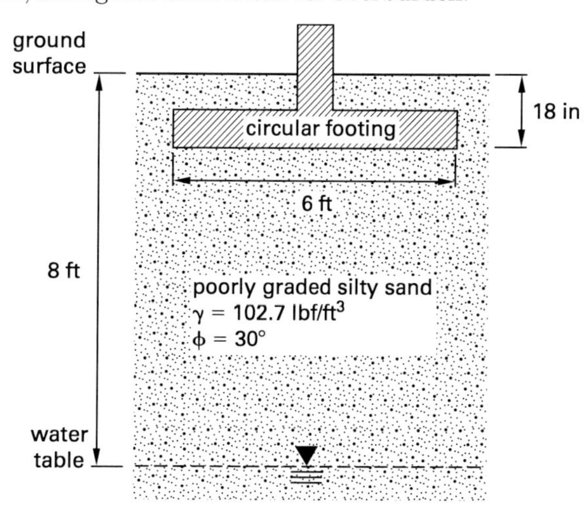

(A) $2600\ \text{lbf/ft}^2$

(B) $3900\ \text{lbf/ft}^2$

(C) $4800\ \text{lbf/ft}^2$

(D) $7700\ \text{lbf/ft}^2$

Hint: The depth of influence due to footing load extends to a depth approximately equivalent to the footing width or diameter.

PROBLEM 3

For a warehouse with large bay spaces, the interior footings are required to support a structural dead plus live load of 100 kips per interior column. Square interior footings are designed in the soil profile as shown. Assume the column loading is concentric.

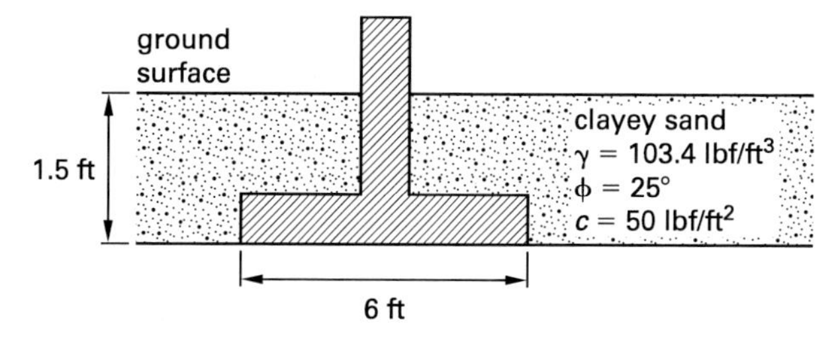

What is most nearly the factor of safety of the given loading with respect to ultimate bearing capacity (using Terzaghi factors)?

(A) 1.6

(B) 2.1

(C) 2.2

(D) 3.1

Hint: Calculate the ultimate bearing capacity, and compare it to the given structural loading.

PROBLEM 4

For the continuous footing shown, approximate the net bearing capacity using the Terzaghi-Meyerhof equation and Terzaghi bearing capacity factors.

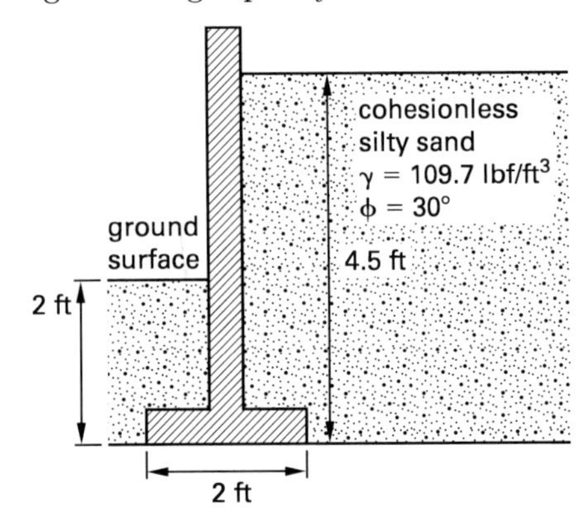

(A) 6880 lbf/ft^2

(B) 7100 lbf/ft^2

(C) 9830 lbf/ft^2

(D) 12,800 lbf/ft^2

Hint: Calculate the ultimate bearing capacity using the bearing depth more likely to fail, and correct it for overburden.

PROBLEM 5

A 7 ft diameter circular footing is constructed 4.5 ft below the ground surface. Given a water table elevation at 2.5 ft below ground surface and the soil parameters shown, what is most nearly the ultimate bearing capacity using the Terzaghi-Meyerhof equation and Meyerhof bearing capacity factors?

$$\gamma_d = 105.9 \text{ lbf/ft}^3$$
$$\phi = 20°$$
$$c = 200 \text{ lbf/ft}^2$$
$$w_{\text{above WT}} = 2\%$$
$$w_{\text{below WT}} = 19\%$$

(A) 5860 lbf/ft^2

(B) 6250 lbf/ft^2

(C) 6700 lbf/ft^2

(D) 8080 lbf/ft^2

Hint: Consider the presence of groundwater within the foundation subgrade, and make the appropriate adjustments to the ultimate bearing capacity equation for buoyancy effects.

PROBLEM 6

Approximate the allowable bearing pressure per unit length of wall of a shallow continuous footing in the soil profile shown. Ignoring correction for overburden, use a factor of safety equal to 2 with Terzaghi bearing capacity factors.

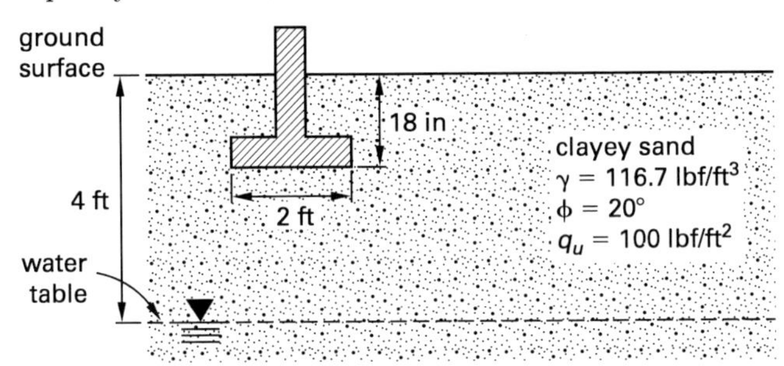

(A) 1220 lbf/ft^2

(B) 1380 lbf/ft^2

(C) 1450 lbf/ft^2

(D) 2760 lbf/ft^2

Hint: Shape factors are not required for a continuous footing.

PROBLEM 7

A square raft foundation sized 50 ft per side is to be constructed with "full compensation" to support an equipment building located at a hydropower generating facility that weighs 2000 tons (including foundation concrete self-weight). The soil consists of lean clay and silt with a moist unit weight of 114.3 lbf/ft^3, water content of 18%, angle of internal friction of 15°, and unconfined compressive strength of 134 lbf/ft^2. No water table is present within 50 ft below the ground surface. What is most nearly the bearing depth?

(A) 7.0 ft

(B) 12 ft

(C) 14 ft

(D) 17 ft

Hint: Full compensation is achieved when the quantity of excavated soil that is equivalent to the overburden pressure is equal to the actual raft contact pressure.

PROBLEM 8

A square mat-and-pedestal foundation is to be constructed at a depth of 10 ft below the ground surface at a site with the soil profile and loading conditions shown. Vertical load includes live plus dead load and foundation self-weight.

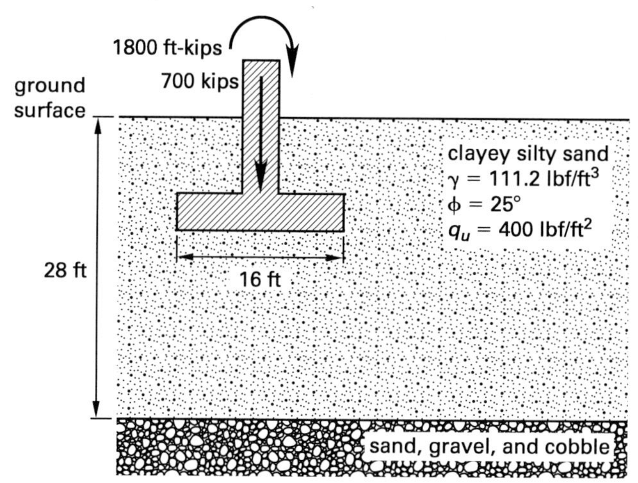

If a factor of safety of 3 is used for allowable bearing pressure, what is most nearly the ratio of the actual soil contact pressure with respect to the allowable bearing pressure?

(A) 1.6

(B) 1.7

(C) 2.0

(D) 3.2

Hint: Determine the eccentricity and equivalent width due to the overturning moment.

PROBLEM 9

Design the width, B, of a shallow continuous (wall) footing to be constructed in the soil profile shown. The design is required to support a structural load of 40 kips/ft with a factor of safety equal to 2. Use Meyerhof bearing capacity factors, ignore the footing self-weight, and correct for overburden.

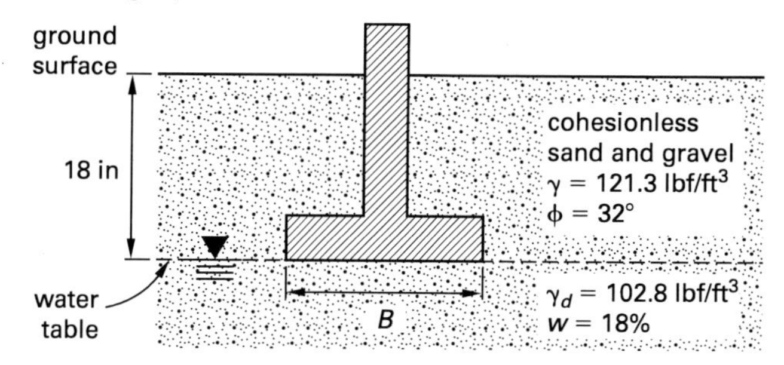

(A) 4 ft

(B) 5 ft

(C) 7 ft

(D) 8 ft

Hint: The groundwater is located at the same level as the base of the proposed footing.

PROBLEM 10

Design the diameter of a shallow footing to be constructed in the soil profile shown. The design is required to support a structural load of 25 kips with a factor of safety equal to 2. Use Terzaghi bearing capacity factors, and ignore the footing self-weight. Ignore correction for overburden.

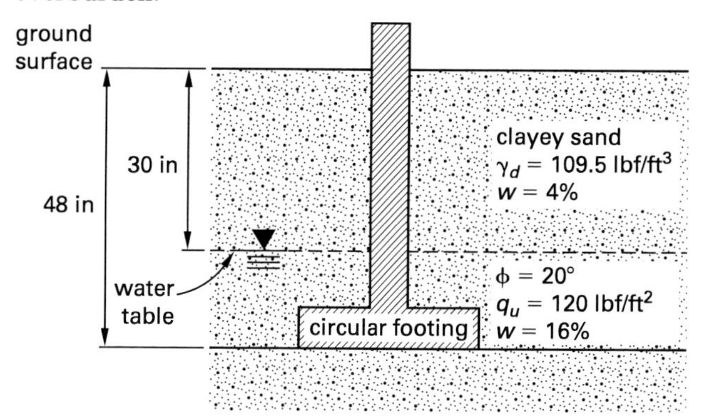

(A) 3.0 ft

(B) 4.0 ft

(C) 5.5 ft

(D) 7.0 ft

Hint: The size of the proposed footing can be determined by using the ultimate bearing capacity equation with the actual soil contact pressure.

PROBLEM 11

For the circular footing shown, a net dead plus live column load of 60 kips and an allowable bearing capacity of 3554 lbf/ft^2 is given.

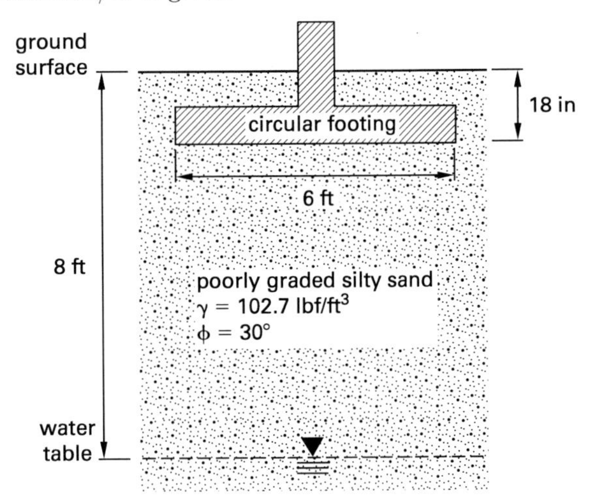

Approximate the actual factor of safety against a bearing capacity failure with respect to the allowable bearing capacity.

(A) 0.059

(B) 0.59

(C) 1.2

(D) 1.7

Hint: Determine the actual load on the footing, and divide the allowable bearing capacity by that actual contact pressure.

PROBLEM 12

A small square mat foundation sized 12 ft is to be constructed at a depth of 2 ft on compacted clayey sand to support a structural load (including self-weight) of 50 tons. The soil consists of a compacted fill with a total unit weight of 113.5 lbf/ft^3, water content of 13%, and angle of internal friction of 25°. Using a factor of safety of 2 and the Terzaghi bearing capacity factors, approximate the ratio between the actual contact pressure and the allowable bearing capacity.

(A) 5.0

(B) 6.0

(C) 7.0

(D) 12

Hint: Determine the net bearing capacity, and compare to the actual soil contact pressure.

SETTLEMENT

PROBLEM 13

Given the soil profile shown, the groundwater location may be assumed to remain constant. Laboratory tests indicate the clay layer is normally consolidated.

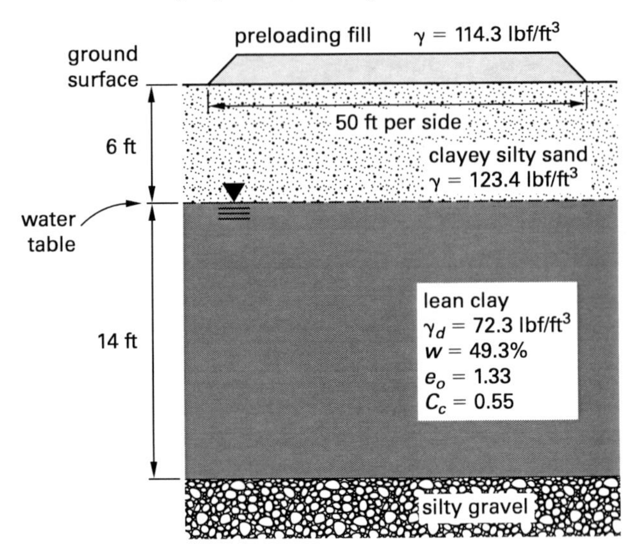

Approximate the settlement of a large fill 10 ft high placed on the original ground surface as shown. Correction for stress dissipation can be ignored.

(A) 1.0 in

(B) 6.3 in

(C) 11 in

(D) 13 in

Hint: To calculate consolidation settlement, the change in surcharge pressure due to the fill is required.

PROBLEM 14

In the soil profile shown, the clay layer is normally consolidated, and the groundwater location may be assumed to remain constant.

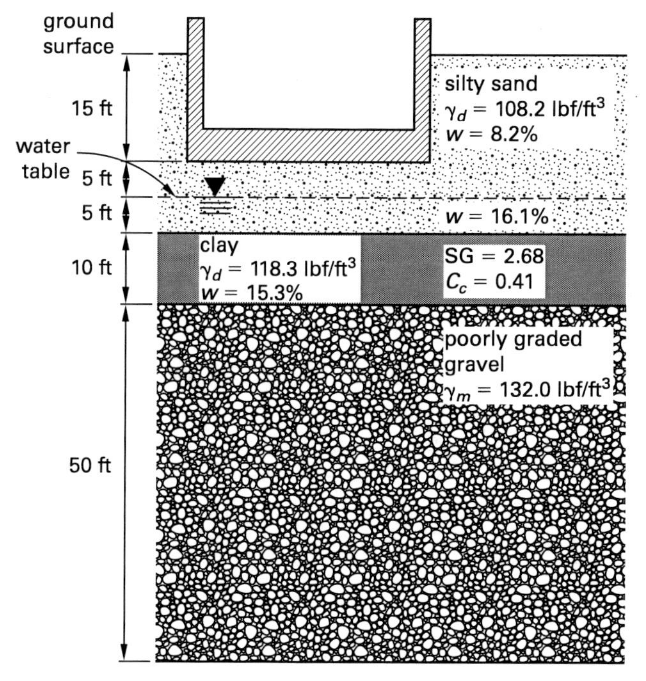

Determine the approximate settlement of a large raft foundation, 50 ft by 50 ft, with a uniform loading of 4000 lbf/ft^2 as shown.

(A) 1 in

(B) 4 in

(C) 7 in

(D) 8 in

Hint: To calculate consolidation settlement, a change in surcharge pressure due to the raft foundation is required.

PROBLEM 15

A 10 ft thick fill embankment of very large extent is to be placed over a clay layer as shown, while the groundwater table is to be simultaneously lowered to 12 ft below ground surface. The clay is saturated with a specific gravity of 2.69 and average water content of 30%, and the compression index was determined to be 0.28.

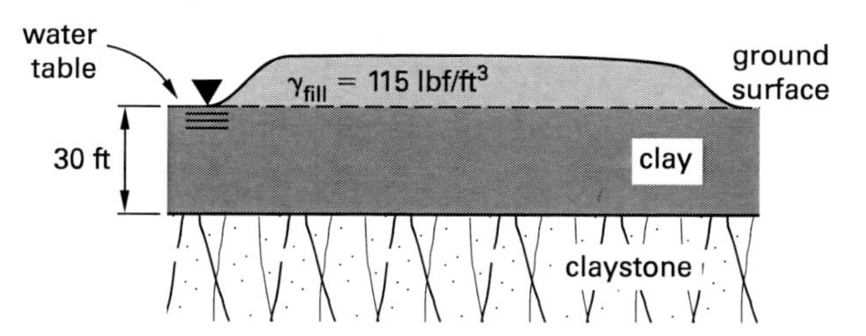

Assuming the clay layer is normally consolidated, determine the approximate settlement of the fill.

(A) 2.0 in

(B) 11 in

(C) 13 in

(D) 28 in

Hint: To calculate consolidation settlement, determine the total unit weight of the clay layer.

VERTICAL STRESS DISTRIBUTION

PROBLEM 16

An "L" shaped load-bearing region is shown. The soil contact pressure of the region is 4000 lbf/ft^2.

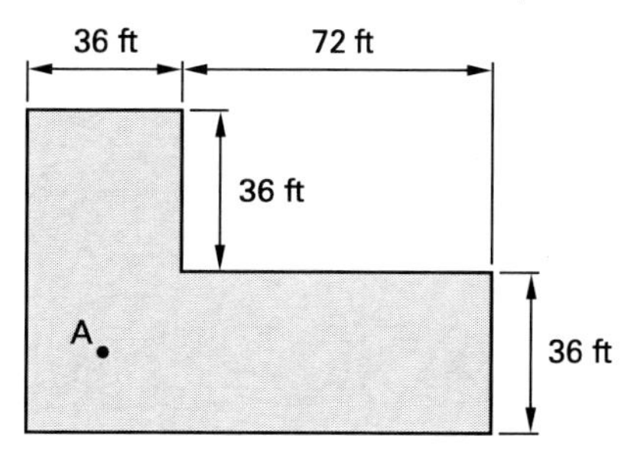

Using Boussinesq stress contours for pressure distribution, approximate the change in total stress at a depth of 18 ft below the ground surface at point A.

(A) 820 lbf/ft^2

(B) 2800 lbf/ft^2

(C) 3300 lbf/ft^2

(D) 4000 lbf/ft^2

Hint: Use superposition and break up the loaded area into smaller squares.

SOLUTION 1

The Terzaghi-Meyerhof bearing capacity equation is

$$q_{ult} = \tfrac{1}{2}\gamma B N_\gamma + cN_c + (p_q + \gamma D_f)N_q$$

The Terzaghi bearing capacity factors are N_γ, N_c, and N_q. p_q is an additional surface surcharge where applicable. For $\phi = 25°$, the following bearing capacity factors were obtained from published tables of Terzaghi's bearing capacity factors.

$$N_\gamma = 9.7$$
$$N_c = 25.1$$
$$N_q = 12.7$$

The bearing capacity can be calculated as

$$\begin{aligned} q_{ult} &= \tfrac{1}{2}\gamma B N_\gamma + cN_c + (p_q + \gamma D_f)N_q \\ &= \left(\frac{1}{2}\right)\left(111.3\ \frac{\text{lbf}}{\text{ft}^3}\right)(10\ \text{ft})(9.7) + \left(275\ \frac{\text{lbf}}{\text{ft}^2}\right)(25.1) \\ &\quad + \left(0 + \left(111.3\ \frac{\text{lbf}}{\text{ft}^3}\right)(36\ \text{in})\left(\frac{1\ \text{ft}}{12\ \text{in}}\right)\right)(12.7) \\ &= 16{,}541\ \text{lbf/ft}^2 \quad (16{,}500\ \text{lbf/ft}^2) \end{aligned}$$

Note that p_q is zero because there is no additional surcharge.

The answer is (C).

Why Other Options Are Wrong

(A) This solution is made incorrect by applying a typical factor of safety, F, equal to 2. This is not the requested solution; the ultimate bearing capacity has been requested, not the allowable bearing capacity with a factor of safety.

(B) This solution mistakenly converts the unit weight to a buoyant unit weight in both terms where unit weight is used. The water table is located too deep for consideration of buoyancy effects.

(D) This solution neglects to multiply the first term by the factor of one half.

SOLUTION 2

The N_c and N_γ factors in the Terzaghi-Meyerhof bearing capacity equation can be assigned S_c and S_γ (shape factors).

$$q_{ult} = \tfrac{1}{2}\gamma B S_\gamma N_\gamma + cS_cN_c + (p_q + \gamma D_f)N_q$$

For footings in sand, the cohesion value, c, is zero. Because additional surface surcharge is not present, the bearing capacity equation is simplified.

$$q_{ult} = \tfrac{1}{2}\gamma B S_\gamma N_\gamma + \gamma D_f N_q$$

The factors N_γ and N_q for $\phi = 30°$ are obtained from published values of Terzaghi bearing capacity factors.

$$N_\gamma = 19.7$$
$$N_q = 22.5$$

The ultimate bearing capacity can be calculated using the appropriate shape factor. S_γ is 0.7 for a circular footing.

$$\begin{aligned} q_{ult} &= \tfrac{1}{2}\gamma B S_\gamma N_\gamma + \gamma D_f N_q \\ &= \left(\frac{1}{2}\right)\left(102.7\ \frac{\text{lbf}}{\text{ft}^3}\right)(6\ \text{ft})(0.7)(19.7) \\ &\quad + \left(102.7\ \frac{\text{lbf}}{\text{ft}^3}\right)(18\ \text{in})\left(\frac{1\ \text{ft}}{12\ \text{in}}\right)(22.5) \\ &= 7715\ \text{lbf/ft}^2 \end{aligned}$$

Using a factor of safety of 2,

$$\begin{aligned} q_a &= \frac{q_{ult}}{F} = \frac{7715\ \frac{\text{lbf}}{\text{ft}^2}}{2} \\ &= 3858\ \text{lbf/ft}^2 \quad (3900\ \text{lbf/ft}^2) \end{aligned}$$

The answer is (B).

Why Other Options Are Wrong

(A) This incorrect solution assumes the water table causes influence from the location shown in the problem statement. The incorrect value is obtained by adjusting for buoyancy effects in the first term of the equation and using a factor of safety equal to 2. Although the groundwater is relatively shallow, its location is not within the influence zone equivalent to the distance of one footing diameter.

(C) This incorrect solution neglects to apply a shape factor for a circular footing.

(D) This incorrect solution calculates the ultimate bearing capacity. It fails to divide by the given factor of safety of 2 for determination of the allowable bearing capacity.

SOLUTION 3

Calculate the ultimate bearing capacity using the Terzaghi-Meyerhof equation (including shape factors) and Terzaghi bearing capacity factors.

$$q_{\text{ult}} = \tfrac{1}{2}\gamma B N_\gamma S_\gamma + c N_c S_c + (p_q + \gamma D_f) N_q$$

The Terzaghi bearing capacity factors are N_γ, N_c, and N_q. p_q is an additional surface surcharge that is not required for this problem. For $\phi = 25°$, the following bearing capacity factors were obtained from published tables.

$$N_\gamma = 9.7$$
$$N_c = 25.1$$
$$N_q = 12.7$$

The shape factors, S_γ and S_c, were obtained from published tables for square footings.

$$S_\gamma = 0.85$$
$$S_c = 1.25$$

The bearing capacity can be calculated as

$$\begin{aligned} q_{\text{ult}} &= \tfrac{1}{2}\gamma B N_\gamma S_\gamma + c N_c S_c + (p_q + \gamma D_f) N_q \\ &= \left(\frac{1}{2}\right)\left(103.4\ \frac{\text{lbf}}{\text{ft}^3}\right)(6\ \text{ft})(9.7)(0.85) \\ &\quad + \left(50\ \frac{\text{lbf}}{\text{ft}^2}\right)(25.1)(1.25) \\ &\quad + \left(0 + \left(103.4\ \frac{\text{lbf}}{\text{ft}^3}\right)(1.5\ \text{ft})\right)(12.7) \\ &= 6096\ \text{lbf/ft}^2 \end{aligned}$$

Note that p_q is zero because there is no additional surcharge.

Calculate the actual soil contact pressure under the footing due to the applied column load.

$$\begin{aligned} p_{\text{actual}} &= \frac{P_{\text{column}}}{A_{\text{footing}}} \\ &= \frac{100\ \text{kips}}{(6\ \text{ft})(6\ \text{ft})} \\ &= 2.778\ \text{kips/ft}^2 \quad (2778\ \text{lbf/ft}^2) \end{aligned}$$

Calculate the factor of safety of the given column loading with respect to the determined ultimate bearing capacity.

$$\begin{aligned} F &= \frac{q_{\text{ult}}}{p_{\text{actual}}} = \frac{6096\ \frac{\text{lbf}}{\text{ft}^2}}{2778\ \frac{\text{lbf}}{\text{ft}^2}} \\ &= 2.2 \end{aligned}$$

The answer is (C).

Why Other Options Are Wrong

(A) Failure to include the cohesion term in the ultimate bearing capacity equation and using the wrong ultimate value to determine the factor of safety results in this incorrect answer.

(B) This solution results from erroneously converting the ultimate bearing capacity to a net bearing capacity and finding the factor of safety with respect to the net bearing capacity.

(D) This solution fails to multiply the first term in the ultimate bearing capacity equation by one half and uses an incorrect value to find the factor of safety.

SOLUTION 4

Using the bearing depth more likely to fail, 2 ft, calculate the ultimate bearing capacity using the Terzaghi-Meyerhof equation and Terzaghi bearing capacity factors.

$$q_{\text{ult}} = \tfrac{1}{2}\gamma B N_\gamma + c N_c + (p_q + \gamma D_f) N_q$$

The Terzaghi bearing capacity factors are N_γ, N_c, and N_q. The cohesion term including N_c is not required, since the given soil is cohesionless. p_q is an additional surface surcharge that is not required for this problem. For $\phi = 30°$, the following bearing capacity factors were obtained from published tables.

$$N_\gamma = 19.7$$
$$N_q = 22.5$$

The ultimate bearing capacity can be calculated.

$$\begin{aligned} q_{\text{ult}} &= \tfrac{1}{2}\gamma B N_\gamma + (p_q + \gamma D_f) N_q \\ &= \left(\frac{1}{2}\right)\left(109.7\ \frac{\text{lbf}}{\text{ft}^3}\right)(2\ \text{ft})(19.7) \\ &\quad + \left(0 + \left(109.7\ \frac{\text{lbf}}{\text{ft}^3}\right)(2\ \text{ft})\right)(22.5) \\ &= 7098\ \text{lbf/ft}^2 \end{aligned}$$

p_q is zero because there is no additional surcharge.

The net bearing capacity can be calculated.

$$\begin{aligned} q_{\text{net}} &= q_{\text{ult}} - \gamma D_f \\ &= 7098 \ \frac{\text{lbf}}{\text{ft}^2} - \left(109.7 \ \frac{\text{lbf}}{\text{ft}^3}\right)(2 \text{ ft}) \\ &= 6879 \text{ lbf/ft}^2 \quad (6880 \text{ lbf/ft}^2) \end{aligned}$$

The answer is (A).

Why Other Options Are Wrong

(B) This incorrect solution is obtained by calculating the ultimate bearing capacity for the footing and neglecting to correct for overburden.

(C) This incorrect solution is obtained by using the average embedment value for the footing.

(D) This incorrect solution is obtained by using the deepest embedment value for the footing.

SOLUTION 5

Calculate the ultimate bearing capacity with adjustment for buoyancy using the Terzaghi-Meyerhof equation (including shape factors) and Meyerhof bearing capacity factors.

$$\begin{aligned} q_{\text{ult}} &= \tfrac{1}{2}\gamma_d B N_\gamma S_\gamma + c N_c S_c \\ &\quad + \left(p_q + \gamma_d D_f + \gamma_w (D_w - D_f)\right) N_q \\ &= \tfrac{1}{2}\left(\gamma_d (1 + w_{\text{below WT}}) - \gamma_w\right) \\ &\quad \times B N_\gamma S_\gamma + c N_c S_c \\ &\quad + \left(p_q + \gamma_d D_f + \gamma_w (D_w - D_f)\right) N_q \end{aligned}$$

The Meyerhof bearing capacity factors are N_γ, N_c, and N_q. p_q is an additional surface surcharge that is not required for this problem. For $\phi = 20°$, the following bearing capacity factors were obtained from published tables.

$$\begin{aligned} N_\gamma &= 2.9 \\ N_c &= 14.8 \\ N_q &= 6.4 \end{aligned}$$

The shape factors, S_γ and S_c, were obtained from published tables for circular footings.

$$\begin{aligned} S_\gamma &= 0.7 \\ S_c &= 1.2 \end{aligned}$$

The ultimate bearing capacity with adjustment for buoyancy can be calculated.

$$\begin{aligned} q_{\text{ult}} &= \tfrac{1}{2}\left(\gamma_d (1 + w_{\text{below WT}}) - \gamma_w\right) \\ &\quad \times B N_\gamma S_\gamma + c N_c S_c \\ &\quad + \left(p_q + \gamma_d D_f + \gamma_w (D_w - D_f)\right) N_q \\ &= \left(\frac{1}{2}\right)\left(\left(105.9 \ \frac{\text{lbf}}{\text{ft}^3}\right)(1 + 0.19) - 62.4 \ \frac{\text{lbf}}{\text{ft}^3}\right) \\ &\quad \times (7 \text{ ft})(2.9)(0.7) + \left(200 \ \frac{\text{lbf}}{\text{ft}^2}\right)(14.8)(1.2) \\ &\quad + \left(\begin{array}{l} 0 + \left(105.9 \ \frac{\text{lbf}}{\text{ft}^3}\right)(4.5 \text{ ft}) \\ \quad + \left(62.4 \ \frac{\text{lbf}}{\text{ft}^3}\right)(2.5 \text{ ft} - 4.5 \text{ ft}) \end{array}\right)(6.4) \\ &= 6256 \text{ lbf/ft}^2 \quad (6250 \text{ lbf/ft}^2) \end{aligned}$$

p_q is zero because there is no additional surcharge.

The answer is (B).

Why Other Options Are Wrong

(A) This incorrect solution is obtained by omitting the shape factors in the first and second terms of the ultimate bearing capacity equation.

(C) This incorrect solution is obtained by omitting the buoyancy correction to the first term of the ultimate bearing capacity equation.

(D) This incorrect solution is obtained by using the ultimate bearing capacity equation without correcting for buoyancy in the first term or the third term.

SOLUTION 6

The effect of groundwater on bearing capacity can be considered negligible since depth of stress influence can be estimated from the footing width. The Terzaghi-Meyerhof equation for ultimate bearing capacity can be used without modification for buoyancy due to the groundwater location. Correction for overburden can be ignored since the depth of the footing is relatively shallow.

$$q_{\text{ult}} = \tfrac{1}{2}\gamma B N_\gamma + c N_c + (p_q + \gamma D_f) N_q$$

The factors N_c, N_γ, and N_q for $\phi = 20°$ are obtained from published tables.

$$\begin{aligned} N_c &= 17.7 \\ N_\gamma &= 5 \\ N_q &= 7.4 \end{aligned}$$

Noting that the cohesion is half the unconfined compressive strength value, the ultimate bearing capacity can be calculated for a continuous footing.

$$c = \frac{q_u}{2} = \frac{100\ \frac{\text{lbf}}{\text{ft}^2}}{2} = 50\ \text{lbf/ft}^2$$

$$\begin{aligned} q_{\text{ult}} &= \tfrac{1}{2}\gamma BN_\gamma + cN_c + (p_q + \gamma D_f)N_q \\ &= \left(\frac{1}{2}\right)\left(116.7\ \frac{\text{lbf}}{\text{ft}^3}\right)(2\ \text{ft})(5) \\ &\quad + \left(50\ \frac{\text{lbf}}{\text{ft}^2}\right)(17.7) + \left(0 + \left(116.7\ \frac{\text{lbf}}{\text{ft}^3}\right)(18\ \text{in})\right. \\ &\quad \times \left.\left(\frac{1\ \text{ft}}{12\ \text{in}}\right)\right)(7.4) \\ &= 2764\ \text{lbf/ft}^2 \end{aligned}$$

Using a factor of safety of 2,

$$\begin{aligned} q_a &= \frac{q_{\text{ult}}}{F} \\ &= \frac{2764\ \frac{\text{lbf}}{\text{ft}^2}}{2} \\ &= 1382\ \text{lbf/ft per ft of wall length} \quad (1380\ \text{lbf/ft}^2) \end{aligned}$$

The answer is (B).

Why Other Options Are Wrong

(A) This solution is incorrect because it assumes the water table causes influence from the location shown in the problem statement. Although the groundwater is relatively shallow, its location is not within the influence zone, typically taken to be a distance between the base of the footing and the water table, or equivalent to the width of the footing. The incorrect value is obtained by adjusting for buoyancy effects in the first term of the equation and using the factor of safety equal to 2.

(C) This solution mistakenly applies a shape factor for a square footing.

(D) This solution fails to divide by the given factor of safety of 2 for determination of the allowable bearing capacity.

SOLUTION 7

The raft foundation is to be fully compensated. Therefore, the structural load is equivalent to the weight of soil to be excavated.

$$\frac{P_{\text{load}}}{A_{\text{raft}}} = \gamma D_f$$

Solve for the required bearing depth.

$$D_f = \frac{\frac{P_{\text{load}}}{A_{\text{raft}}}}{\gamma} = \frac{\frac{(2000\ \text{tons})\left(2000\ \frac{\text{lbf}}{\text{ton}}\right)}{(50\ \text{ft})(50\ \text{ft})}}{114.3\ \frac{\text{lbf}}{\text{ft}^3}} = 14\ \text{ft}$$

The calculation of bearing capacity is not required because this is a fully compensated case.

The answer is (C).

Why Other Options Are Wrong

(A) This solution fails to convert the value of tons to pounds prior to the calculation of bearing depth.

(B) This solution mistakes the given total unit weight for a dry unit weight and calculates a new total unit weight value.

(D) This solution is incorrect because it converts the unit weight value to a dry unit weight value. The total unit weight given should be used, since moisture adds to the overburden weight.

SOLUTION 8

An eccentricity is created due to overturning moment in the direction shown. The location of the eccentricity can be calculated.

$$\begin{aligned} \varepsilon &= \frac{M}{P} = \frac{1800\ \text{ft-kips}}{700\ \text{kips}} \\ &= 2.6\ \text{ft} \\ \frac{L}{6} &= \frac{16\ \text{ft}}{6} = 2.7\ \text{ft} \end{aligned}$$

Compare the distance of the eccentricity from the center of the footing to determine the equation for soil contact pressure criteria.

$$\begin{aligned} 2.6\ \text{ft} &< 2.7\ \text{ft} \\ \varepsilon &< \frac{L}{6} \end{aligned}$$

The eccentricity is located inside the middle third of the footing width; therefore, the soil contact pressure is distributed according to the equation

$$p_{\text{max}}, p_{\text{min}} = \left(\frac{P}{BL}\right)\left(1 \pm \frac{6\varepsilon}{B}\right)$$

The maximum and minimum values are

$$p_{\text{max}} = \left(\frac{P}{BL}\right)\left(1 + \frac{6\varepsilon}{B}\right)$$
$$= \left(\frac{7.0 \times 10^5 \text{ lbf}}{(16 \text{ ft})(16 \text{ ft})}\right)\left(1 + \frac{(6)(2.6 \text{ ft})}{16 \text{ ft}}\right)$$
$$= 5400 \text{ lbf/ft}^2$$
$$p_{\text{min}} = \left(\frac{P}{BL}\right)\left(1 - \frac{6\varepsilon}{B}\right)$$
$$= \left(\frac{7.0 \times 10^5 \text{ lbf}}{(16 \text{ ft})(16 \text{ ft})}\right)\left(1 - \frac{(6)(2.6 \text{ ft})}{16 \text{ ft}}\right)$$
$$= 68 \text{ lbf/ft}^2$$

The minimum result is positive. Therefore, the entire footing is in compression, and the contact pressure is distributed over the entire footing area. Determine the shape factors, and use the Terzaghi-Meyerhof bearing capacity equation to find the ultimate bearing capacity. The shape factors, S_γ and S_c, are obtained from published tables for square footings.

$$\frac{B}{L} = \frac{16 \text{ ft}}{16 \text{ ft}} = 1$$
$$S_\gamma = 0.85$$
$$S_c = 1.25$$

The following Terzaghi bearing capacity factors were obtained from published tables.

$$N_\gamma = 9.7$$
$$N_c = 25.1$$
$$N_q = 12.7$$

The bearing capacity can be calculated using the unconfined compressive strength. However, the value for cohesion is obtained from dividing the unconfined compressive strength by two.

$$c = \frac{q_u}{2} = \frac{400 \frac{\text{lbf}}{\text{ft}^2}}{2} = 200 \text{ lbf/ft}^2$$
$$q_{\text{ult}} = \tfrac{1}{2}\gamma L N_\gamma S_\gamma + cN_c S_c + (p_q + \gamma D_f)N_q$$
$$= \left(\frac{1}{2}\right)\left(111.2 \frac{\text{lbf}}{\text{ft}^3}\right)(16 \text{ ft})(9.7)(0.85)$$
$$+ \left(200 \frac{\text{lbf}}{\text{ft}^2}\right)(25.1)(1.25)$$
$$+ \left(0 + \left(111.2 \frac{\text{lbf}}{\text{ft}^3}\right)(10 \text{ ft})\right)(12.7)$$
$$= 27{,}732 \text{ lbf/ft}^2$$

Although p_q is zero because there is no additional surcharge, the ultimate bearing capacity should be corrected for overburden.

$$q_{\text{net}} = q_{\text{ult}} - \gamma D_f$$
$$= 27{,}732 \frac{\text{lbf}}{\text{ft}^2} - \left(111.2 \frac{\text{lbf}}{\text{ft}^3}\right)(10 \text{ ft})$$
$$= 26{,}620 \text{ lbf/ft}^2$$

The allowable bearing capacity can be determined using the factor of safety.

$$q_a = \frac{q_{\text{net}}}{F} = \frac{26{,}620 \frac{\text{lbf}}{\text{ft}^2}}{3}$$
$$= 8873 \text{ lbf/ft}^2$$

Calculate the ratio of the actual soil contact pressure with respect to the allowable bearing capacity.

$$F = \frac{q_a}{p_{\text{max}}} = \frac{8873 \frac{\text{lbf}}{\text{ft}^2}}{5400 \frac{\text{lbf}}{\text{ft}^2}}$$
$$= 1.6$$

The answer is (A).

Why Other Options Are Wrong

(B) This incorrect solution omits the determination of the net bearing capacity.

(C) This incorrect solution uses the full value of the unconfined compressive strength (instead of dividing it in half) to find cohesion in calculating the ultimate bearing capacity.

(D) This incorrect solution ignores the determination of the soil contact pressure by considering the location of the eccentricity. The actual contact pressure is determined by dividing the vertical load by the footing area.

SOLUTION 9

The effect of groundwater on bearing capacity should be considered in the design. The Terzaghi-Meyerhof equation for ultimate bearing capacity can be used with modification of the first term for buoyancy due to the groundwater location. Although the depth of the footing is relatively shallow, correction for overburden is to be considered.

$$q_{\text{ult}} = \tfrac{1}{2}\gamma_d B N_\gamma + (p_q + \gamma D_f)N_q$$

The Meyerhof factors, N_γ and N_q, for $\phi = 32°$ are obtained from published tables.

$$N_\gamma = 22.0$$
$$N_q = 23.2$$

The ultimate bearing capacity can be written for a continuous footing with the unknown width, along with consideration for the water table being located at the base of the footing.

$$\begin{aligned} q_{\text{ult}} &= \tfrac{1}{2}\gamma_d B N_\gamma + (p_q + \gamma D_f) N_q \\ &= \left(\frac{1}{2}\right)\left(\left(102.8\ \frac{\text{lbf}}{\text{ft}^3}\right)(1+0.18) - 62.4\ \frac{\text{lbf}}{\text{ft}^3}\right)B(22.0) \\ &\quad + \left(0 + \left(121.3\ \frac{\text{lbf}}{\text{ft}^3}\right)(18\ \text{in})\left(\frac{1\ \text{ft}}{12\ \text{in}}\right)\right)(23.2) \\ &= 648B\ \text{lbf/ft}^3 + 4221\ \text{lbf/ft}^2 \end{aligned}$$

The net bearing capacity can be calculated.

$$\begin{aligned} q_{\text{net}} &= q_{\text{ult}} - \gamma D_f \\ &= 648B\ \frac{\text{lbf}}{\text{ft}^3} + 4221\ \frac{\text{lbf}}{\text{ft}^2} - \left(121.3\ \frac{\text{lbf}}{\text{ft}^3}\right)(18\ \text{in})\left(\frac{1\ \text{ft}}{12\ \text{in}}\right) \\ &= 648B\ \text{lbf/ft}^3 + 4039\ \text{lbf/ft}^2 \end{aligned}$$

Using a factor of safety of 2, calculate the allowable bearing capacity.

$$\begin{aligned} q_a &= \frac{q_{\text{net}}}{F} \\ &= \frac{648B\ \frac{\text{lbf}}{\text{ft}^3} + 4039\ \frac{\text{lbf}}{\text{ft}^2}}{2} \\ &= 324B\ \text{lbf/ft}^3 + 2020\ \text{lbf/ft}^2 \end{aligned}$$

Express the structural load in terms of the footing width.

$$\begin{aligned} p_{\text{actual}} &= \frac{\text{given column load}}{\text{footing width}} \\ &= \frac{40\ \frac{\text{kips}}{\text{ft}}}{B_{\text{footing}}} \end{aligned}$$

Equate the two expressions and solve for the width.

$$B_{\text{footing}} = B$$
$$\frac{40\ \frac{\text{kips}}{\text{ft}}}{B_{\text{footing}}} = 324B\ \frac{\text{lbf}}{\text{ft}^3} + 2020\ \frac{\text{lbf}}{\text{ft}^2}$$
$$0 = 324B^2\ \frac{\text{lbf}}{\text{ft}^3} + 2020B\ \frac{\text{lbf}}{\text{ft}^2} - 40{,}000\ \frac{\text{lbf}}{\text{ft}}$$

Trial-and-error iteration may be used to solve for the width but, recognizing that this is a second-degree polynomial, the use of the quadratic equation is another option.

$$x_1, x_2 = \frac{-b \pm \sqrt{b^2 - 4ac}}{2a}$$

Substitute into the equation using $a = 324$, $b = 2020$, and $c = -40{,}000$.

$$\begin{aligned} x_1 &= \frac{-b + \sqrt{b^2 - 4ac}}{2a} \\ &= \frac{-2020 + \sqrt{(2020)^2 - (4)(324)(-40{,}000)}}{(2)(324)} \\ &= 8.4 \quad (8\ \text{ft}) \end{aligned}$$

$$\begin{aligned} x_2 &= \frac{-b - \sqrt{b^2 - 4ac}}{2a} \\ &= \frac{-2020 - \sqrt{(2020)^2 - (4)(324)(-40{,}000)}}{(2)(324)} \\ &= -15 \quad [\text{disregard this root}] \end{aligned}$$

The answer is (D).

Why Other Options Are Wrong

(A) This solution fails to divide the ultimate bearing capacity value by the factor of safety of 2.

(B) This solution fails to consider the effect of the water table, resulting in failure to adjust for buoyancy effects in the first term of the equation, but still using the factor of safety equal to 2. The ultimate bearing capacity is written for a continuous footing with the unknown width, but without consideration for the water table being located at the base of the footing.

(C) This solution fails to adjust the given dry density to a buoyant density in the first term of the ultimate bearing capacity equation.

SOLUTION 10

The Terzaghi-Meyerhof equation for ultimate bearing capacity can be solved for the radius using the appropriate shape factors, S_c and S_γ, for a circular footing. The effect of groundwater on bearing capacity is to be considered in the design. Therefore, the ultimate bearing capacity equation should be used with modification to the first and third terms due to the groundwater location.

$$q_{\text{ult}} = \tfrac{1}{2}\gamma_d(2R)S_\gamma N_\gamma + cS_cN_c + \big(p_q + (\gamma_d D_f + \gamma_w(D_w - D_f))\big)N_q$$

The Terzaghi factors, N_c, N_γ, and N_q, for $\phi = 20°$ are obtained from published tables.

$$N_\gamma = 5.0$$
$$N_c = 17.7$$
$$N_q = 7.4$$

The shape factors, S_γ and S_c, are obtained from published tables for a circular footing.

$$S_\gamma = 0.70$$
$$S_c = 1.20$$

The ultimate bearing capacity can be written for a continuous footing with an unknown width and with consideration for the water table being located between the ground surface and the base of the footing. The effective unit weight is being used in the first term. The depth to the water table, 30 in, is 2.5 ft, and the depth to the bottom of the footing, 48 in, is 4 ft.

$$c = \frac{q_u}{2} = \frac{120\ \frac{\text{lbf}}{\text{ft}^2}}{2} = 60\ \text{lbf/ft}^2$$

$$\begin{aligned} q_{\text{ult}} &= \tfrac{1}{2}\gamma_d(2R)S_\gamma N_\gamma + cS_cN_c + \big(p_q + (\gamma_d D_f + \gamma_w(D_w - D_f))\big)N_q \\ &= \left(\frac{1}{2}\right)\left(\left(109.5\ \frac{\text{lbf}}{\text{ft}^3}\right)(1+0.16) - 62.4\ \frac{\text{lbf}}{\text{ft}^3}\right) \\ &\quad \times 2R(0.70)(5.0) + \left(60\ \frac{\text{lbf}}{\text{ft}^2}\right)(1.20)(17.7) \\ &\quad + \left(0 + \left(109.5\ \frac{\text{lbf}}{\text{ft}^3}\right)(4\ \text{ft}) + \left(62.4\ \frac{\text{lbf}}{\text{ft}^3}\right)(2.5\ \text{ft} - 4\ \text{ft})\right)(7.4) \\ &= \left(226.2\ \frac{\text{lbf}}{\text{ft}^3}\right)R + 1274.4\ \frac{\text{lbf}}{\text{ft}^2} + 2548.6\ \frac{\text{lbf}}{\text{ft}^2} \\ &= \left(226.2\ \frac{\text{lbf}}{\text{ft}^3}\right)R + 3823\ \frac{\text{lbf}}{\text{ft}^2} \end{aligned}$$

The net bearing capacity can be disregarded, since correction for overburden can be ignored.

Using a factor of safety of 2, express the allowable bearing capacity in terms of radius.

$$\begin{aligned} q_a &= \frac{q_{\text{ult}}}{F} \\ &= \frac{\left(226.2\ \frac{\text{lbf}}{\text{ft}^3}\right)R + 3823\ \frac{\text{lbf}}{\text{ft}^2}}{2} \\ &= \left(113.1\ \frac{\text{lbf}}{\text{ft}^3}\right)R + 1911.5\ \frac{\text{lbf}}{\text{ft}^2} \end{aligned}$$

Express the actual soil contact pressure in terms of the footing radius.

$$\begin{aligned} p_{\text{actual}} &= \frac{P_{\text{column}}}{A_{\text{footing}}} = \frac{P}{\pi R^2} \\ &= \frac{25\ \text{kips}}{\pi R^2} \end{aligned}$$

Equate the two expressions and solve for the radius by trial-and-error iteration.

$$\begin{aligned} \frac{25\ \text{kips}}{\pi R^2} &= \left(113.1\ \frac{\text{lbf}}{\text{ft}^3}\right)R + 1911.5\ \frac{\text{lbf}}{\text{ft}^2} \\ 25{,}000\ \text{lbf} &= \left(355.3\ \frac{\text{lbf}}{\text{ft}^3}\right)R^3 + \left(6005.2\ \frac{\text{lbf}}{\text{ft}^2}\right)R^2 \end{aligned}$$

Commence with a reasonable value for the radius, such as 2 ft.

$$\begin{aligned} &\left(355.3\ \frac{\text{lbf}}{\text{ft}^3}\right)(2\ \text{ft})^3 + \left(6005.2\ \frac{\text{lbf}}{\text{ft}^2}\right)(2\ \text{ft})^2 \\ &\quad = 26{,}863\ \text{lbf} > 25{,}000\ \text{lbf} \end{aligned}$$

The value of 2 ft for the radius results in a larger than desired value. However, due to the third-degree equation, 1.5 ft results in a value less than 25 kips. Therefore, a 4.0 ft diameter is the correct answer.

The answer is (B).

Why Other Options Are Wrong

(A) This solution fails to divide the ultimate bearing capacity value by the factor of safety of 2. Iteration is still required until a correct value is obtained.

(C) This solution mistakenly applies the factor of safety to both the bearing capacity and the load conditions.

(D) This solution fails to multiply the soil contact expression by π in solving for the radius.

SOLUTION 11

The actual factor of safety against the bearing capacity failure is determined by

$$F_{\text{actual}} = \frac{q_a}{\frac{P}{A}} = \frac{q_a}{\frac{P}{\frac{\pi d^2}{4}}} = \frac{3554 \ \frac{\text{lbf}}{\text{ft}^2}}{\frac{60{,}000 \text{ lbf}}{\frac{\pi(6 \text{ ft})^2}{4}}}$$
$$= 1.68 \quad (1.7)$$

The answer is (D).

Why Other Options Are Wrong

(A) This incorrect solution divides the allowable bearing capacity by the column load without dividing by the footing area first. The units are incorrect.

(B) This incorrect solution is found by dividing the actual applied bearing pressure by the allowable bearing pressure, the reciprocal of the correct answer.

(C) This incorrect solution assumes that the water table causes influence from the location shown in the problem statement. The incorrect value is obtained by adjusting for buoyancy effects in the first term of the ultimate bearing capacity equation, dividing by the factor of safety, and then dividing that incorrect value for allowable bearing capacity by the actual bearing pressure.

SOLUTION 12

The ratio is to be calculated by comparing the allowable bearing capacity (obtained from the ultimate bearing capacity and net bearing capacity) to the actual soil contact pressure. Calculate the ultimate bearing capacity. Assume it is cohesionless, since no value for cohesion is given.

$$q_{\text{ult}} = \tfrac{1}{2}\gamma B N_\gamma S_\gamma + (p_q + \gamma D_f)N_q$$

The factors N_γ and N_q for $\phi = 25°$ are obtained from published tables.

$$N_\gamma = 9.7$$
$$N_q = 12.7$$

The shape factor, S_γ, can be obtained from published tables for square footings.

$$S_\gamma = 0.85$$

The ultimate bearing capacity can be calculated for a square footing using the shape factor.

$$q_{\text{ult}} = \tfrac{1}{2}\gamma B N_\gamma S_\gamma + (p_q + \gamma D_f)N_q$$
$$= \left(\frac{1}{2}\right)\left(113.5 \ \frac{\text{lbf}}{\text{ft}^3}\right)(12 \text{ ft})(9.7)(0.85)$$
$$+ \left(0 + \left(113.5 \ \frac{\text{lbf}}{\text{ft}^3}\right)(2 \text{ ft})\right)(12.7)$$
$$= 8498 \text{ lbf/ft}^2$$

p_q is zero because there is no additional surcharge. The ultimate bearing capacity should be corrected for overburden (approximately the dead weight above the foundation plane and below the ground level).

$$q_{\text{net}} = q_{\text{ult}} - \gamma D_f = 8498 \ \frac{\text{lbf}}{\text{ft}^2} - \left(113.5 \ \frac{\text{lbf}}{\text{ft}^3}\right)(2 \text{ ft})$$
$$= 8271 \text{ lbf/ft}^2$$

The allowable bearing capacity can be determined using the given factor of safety.

$$q_a = \frac{q_{\text{net}}}{F} = \frac{8271 \ \frac{\text{lbf}}{\text{ft}^2}}{2} = 4136 \text{ lbf/ft}^2$$

Calculate the factor of safety of the actual soil contact pressure with respect to the allowable bearing capacity.

$$F = \frac{q_a}{\frac{P_{\text{max}}}{A_{\text{mat}}}} = \frac{4136 \ \frac{\text{lbf}}{\text{ft}^2}}{\frac{(50 \text{ tons})\left(2000 \ \frac{\text{lbf}}{\text{ton}}\right)}{(12 \text{ ft})(12 \text{ ft})}}$$
$$= 6.0$$

The answer is (B).

Why Other Options Are Wrong

(A) This solution mistakenly converts the given total unit weight to a dry unit weight.

(C) This solution fails to apply a shape factor to determine the ultimate bearing capacity for a continuous footing.

(D) This solution fails to divide the net bearing capacity by the given factor of safety for determination of the allowable bearing capacity.

SOLUTION 13

The equation for consolidation settlement can be expressed as

$$S_c = C_c\left(\frac{H_o}{1+e_o}\right)\log_{10}\frac{p_o' + \Delta p_v'}{p_o'}$$

Determine the total increase in surcharge load due to the fill. Stress influence factors can be ignored.

$$\begin{aligned}\Delta p_v' = \gamma H &= \left(114.3\ \frac{\text{lbf}}{\text{ft}^3}\right)(10\ \text{ft})\\ &= 1143\ \text{lbf/ft}^2\end{aligned}$$

As indicated in the problem statement, the depth of the groundwater can be assumed to remain constant. Determine the initial effective stress at the midpoint of the clay layer by first determining the initial total stress halfway through the clay layer and subtracting the pore pressure.

$$\begin{aligned}p_o' &= p_{\text{sand}} + (p_{\text{clay,midpoint}} - \mu)\\ &= \gamma H_{\text{sand}} + (\gamma_d(1+w) - \gamma_w)\left(\frac{H_o}{2}\right)\\ &= \left(123.4\ \frac{\text{lbf}}{\text{ft}^3}\right)(6\ \text{ft})\\ &\quad + \left(\left(72.3\ \frac{\text{lbf}}{\text{ft}^3}\right)(1+0.493) - 62.4\ \frac{\text{lbf}}{\text{ft}^3}\right)\left(\frac{14\ \text{ft}}{2}\right)\\ &= 1059\ \text{lbf/ft}^2\end{aligned}$$

Determine the total expected settlement due to placement of the large fill.

$$\begin{aligned}S_c &= C_c\left(\frac{H_o}{1+e_o}\right)\log_{10}\frac{p_o' + \Delta p_v'}{p_o'}\\ &= (0.55)\left(\frac{14\ \text{ft}}{1+1.33}\right)\\ &\quad \times\left(\log_{10}\frac{1059\ \frac{\text{lbf}}{\text{ft}^2} + 1143\ \frac{\text{lbf}}{\text{ft}^2}}{1059\ \frac{\text{lbf}}{\text{ft}^2}}\right)\left(12\ \frac{\text{in}}{\text{ft}}\right)\\ &= 12.6\ \text{in} \quad (13\ \text{in})\end{aligned}$$

The answer is (D).

Why Other Options Are Wrong

(A) This incorrect solution is actually the approximate settlement value in feet obtained by failing to convert the calculation to inches.

(B) This solution erroneously uses half of the clay layer thickness for determining the settlement of the clay layer.

(C) This solution neglects to subtract pore water pressure when determining the initial effective stress. This value is also low in comparison to the correct answer.

SOLUTION 14

The equation for consolidation settlement can be expressed as

$$S = C_c\left(\frac{H_o}{1+e_o}\right)\log_{10}\frac{p_o' + \Delta p_v'}{p_o'}$$

To determine the change in effective stress due to the total increase in surcharge load, first calculate the initial effective stress at the midpoint location of the clay layer.

$$\begin{aligned}p_o' &= p_{\text{sand,above WT}} + p_{\text{sand,below WT}} + p_{\text{clay,midpoint}} - u\\ &= \gamma_d(1+w_{\text{above WT}})D_{\text{sand,above WT}}\\ &\quad + \gamma_d(1+w_{\text{below WT}})t_{\text{sand,below WT}}\\ &\quad + \gamma_d(1+w_{\text{clay}})\left(\frac{t_{\text{clay}}}{2}\right) - \gamma_w D_w\\ &= \left(108.2\ \frac{\text{lbf}}{\text{ft}^3}\right)(1+0.082)(20\ \text{ft})\\ &\quad + \left(108.2\ \frac{\text{lbf}}{\text{ft}^3}\right)(1+0.161)(5\ \text{ft})\\ &\quad + \left(118.3\ \frac{\text{lbf}}{\text{ft}^3}\right)(1+0.153)\left(\frac{10\ \text{ft}}{2}\right)\\ &\quad - \left(62.4\ \frac{\text{lbf}}{\text{ft}^3}\right)(10\ \text{ft})\\ &= 3028\ \text{lbf/ft}^2\end{aligned}$$

The change in effective stress can be calculated by first finding the net surcharge pressure.

$$\begin{aligned}p_{\text{net,center}} &= p_q - \gamma_t H\\ &= p_q - \gamma_d(1+w_{\text{above WT}})D_f\\ &= 4000\ \frac{\text{lbf}}{\text{ft}^2} - \left(108.2\ \frac{\text{lbf}}{\text{ft}^3}\right)\\ &\quad \times(1+0.082)(15\ \text{ft})\\ &= 2244\ \text{lbf/ft}^2\end{aligned}$$

Then, a Boussinesq influence factor chart is required for determining the change in stress at the midpoint of the clay layer. To use a Boussinesq chart, the stress influence factor corresponding to $(15\ \text{ft}/50\ \text{ft})B = 0.3B$ is required for determining the stress at the midpoint of the clay layer beneath the center point of the raft foundation. The factor is found to be $0.9p$.

$$\begin{aligned}\Delta p_v' &= 0.9p = 0.9p_{\text{net,center}}\\ &= (0.9)\left(2244\ \frac{\text{lbf}}{\text{ft}^2}\right)\\ &= 2020\ \text{lbf/ft}^2\end{aligned}$$

Determine the total expected settlement due to placement of the raft foundation. Calculate the void ratio for use in the settlement equation assuming 100% saturation for the submerged clay.

$$\begin{aligned}e_o &= \frac{w(\text{SG})}{S} = \frac{(0.153)(2.68)}{1}\\ &= 0.41\\ S &= C_c\left(\frac{H_o}{1+e_o}\right)\log_{10}\frac{p_o' + \Delta p_v'}{p_o'}\\ &= (0.41)\left(\frac{10\ \text{ft}}{1+0.41}\right)\log_{10}\left(\frac{3028\ \frac{\text{lbf}}{\text{ft}^2} + 2020\ \frac{\text{lbf}}{\text{ft}^2}}{3028\ \frac{\text{lbf}}{\text{ft}^2}}\right)\\ &\quad\times\left(12\ \frac{\text{in}}{\text{ft}}\right)\\ &= 7.7\ \text{in}\quad (8\ \text{in})\end{aligned}$$

The answer is (D).

Why Other Options Are Wrong

(A) This answer is the approximate settlement value, in feet, obtained by failing to convert the calculation to inches. This result is very low compared to the correct settlement, since the units are typically rounded up and treated as inches.

(B) This solution erroneously uses half of the clay layer thickness for determining the settlement of the clay layer.

(C) Neglecting to subtract pore water pressure when determining the initial effective stress results in this incorrect answer. This value is also low in comparison to the correct answer.

SOLUTION 15

The equation for consolidation settlement is

$$S = C_c\left(\frac{H_o}{1+e_o}\right)\log_{10}\frac{p_o' + \Delta p_v'}{p_o'}$$

To determine the initial effective stress at the midpoint of the clay layer, the average saturated unit weight should be calculated. Use phase relationships to determine the required parameters. Use the water content and the specific gravity values to find the initial void ratio and the total volume corresponding to the saturated unit weight of the clay.

$$\begin{aligned}e_o &= \frac{w(\text{SG})}{S} = \frac{(0.3)(2.69)}{1}\\ &= 0.807\end{aligned}$$

Find the dry density and saturated density of the clay.

$$\begin{aligned}\gamma_d &= \frac{(\text{SG})\gamma_w}{1+e_o} = \frac{(2.69)\left(62.4\ \frac{\text{lbf}}{\text{ft}^3}\right)}{1+0.807}\\ &= 92.9\ \text{lbf/ft}^3\\ \gamma_{\text{sat}} &= \gamma_d(1+w) = \left(92.9\ \frac{\text{lbf}}{\text{ft}^3}\right)(1+0.3)\\ &= 120.8\ \text{lbf/ft}^3\end{aligned}$$

Calculate the initial effective stress at the midpoint of the clay layer.

$$\begin{aligned}p_o' &= p_{\text{clay,midpoint}} - u_o = \gamma_t D_o - \gamma_w D_o\\ &= (\gamma_{\text{sat}} - \gamma_w)D_o\\ &= \left(120.8\ \frac{\text{lbf}}{\text{ft}^3} - 62.4\ \frac{\text{lbf}}{\text{ft}^3}\right)(15\ \text{ft})\\ &= 876\ \text{lbf/ft}^2\end{aligned}$$

Determine the change in effective stress referenced from the midpoint of the clay layer due to the total increase in surcharge load from the fill and a simultaneous change in water table height.

$$\begin{aligned}\Delta p_v' &= p_q + (u_o - u_f) = p_q + (H_{w,o} - H_{w,f})\gamma_w\\ &= p_q + (D_o - D_f)\gamma_w\\ p_q &= \gamma_{\text{fill}}H_{\text{fill}}\\ &= \left(115\ \frac{\text{lbf}}{\text{ft}^3}\right)(10\ \text{ft})\\ &= 1150\ \text{lbf/ft}^2\\ \Delta p_v' &= p_q + (D_o - D_f)\gamma_w\\ &= 1150\ \frac{\text{lbf}}{\text{ft}^2} + (15\ \text{ft} - 3\ \text{ft})\left(62.4\ \frac{\text{lbf}}{\text{ft}^3}\right)\\ &= 1899\ \text{lbf/ft}^2\end{aligned}$$

Calculate the total expected settlement.

$$S = C_c\left(\frac{H_o}{1+e_o}\right)\log_{10}\frac{p_o' + \Delta p_v'}{p_o'}$$

$$= (0.28)\left(\frac{30 \text{ ft}}{1+0.807}\right)\log_{10}\left(\frac{876 \frac{\text{lbf}}{\text{ft}^2} + 1899 \frac{\text{lbf}}{\text{ft}^2}}{876 \frac{\text{lbf}}{\text{ft}^2}}\right) \times \left(12 \frac{\text{in}}{\text{ft}}\right)$$

$$= 28 \text{ in}$$

The answer is (D).

Why Other Options Are Wrong

(A) This solution is the approximate settlement value, in feet, obtained by failing to convert the calculation to inches. The units are not correct.

(B) This incorrect solution uses half of the clay layer thickness for determining the settlement of the clay layer.

(C) This solution neglects to subtract pore water pressure when determining the initial effective stress.

SOLUTION 16

To use the Boussinesq stress contour chart for square shapes, divide the loaded area into smaller squares with similar sides as shown.

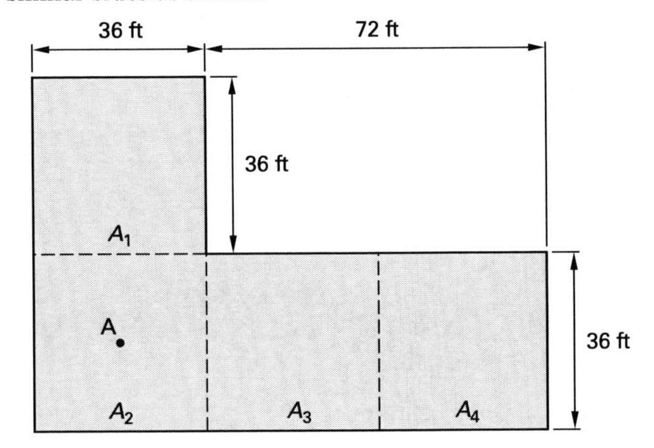

The increase in total stress at a depth of 18 ft for each area can be determined using the Boussinesq stress contour chart. For area A_1, determine the properties required to find the influence factor.

$$B = 36 \text{ ft}$$
$$D = 18 \text{ ft} = 0.5B$$

Point A is located at a perpendicular distance equivalent to the width of area A_1 away from the center of area A_1. The influence factor corresponds to the stress contour located at $0.5B$ down and B over from the center of the foundation sketch depicted on the chart, and the value can be read from the chart.

$$p_{A_1,18} = 0.06p_{\text{applied}}$$
$$p_{\text{applied}} = 4000 \text{ lbf/ft}^2$$
$$p_{A_1,18} = 0.06p_{\text{applied}} = (0.06)\left(4000 \frac{\text{lbf}}{\text{ft}^2}\right)$$
$$= 240 \text{ lbf/ft}^2$$

Similarly, the additional vertical stress for the remaining areas can be determined from the chart.

$$p_{A_2,18} = 0.7p_{\text{applied}} = (0.7)\left(4000 \frac{\text{lbf}}{\text{ft}^2}\right) = 2800 \text{ lbf/ft}^2$$

$$p_{A_3,18} = p_{A_1,18} = 240 \text{ lbf/ft}^2$$

$$p_{A_4,18} = 0.0025p_{\text{applied}} = (0.0025)\left(4000 \frac{\text{lbf}}{\text{ft}^2}\right) = 10 \text{ lbf/ft}^2$$

Calculate the total increase in vertical stress at 18 ft below point A.

$$p_{18} = p_{A_1,18} + p_{A_2,18} + p_{A_3,18} + p_{A_4,18}$$
$$= 240 \frac{\text{lbf}}{\text{ft}^2} + 2800 \frac{\text{lbf}}{\text{ft}^2} + 240 \frac{\text{lbf}}{\text{ft}^2} + 10 \frac{\text{lbf}}{\text{ft}^2}$$
$$= 3290 \text{ lbf/ft}^2 \quad (3300 \text{ lbf/ft}^2)$$

The answer is (C).

Why Other Options Are Wrong

(A) This answer wrongly assumes that the distributed load for each square area is one fourth of the actual load value, or 1000 lbf/ft^2, rather than 4000 lbf/ft^2.

(B) This incorrect answer is obtained by determining the change in vertical pressure due only to the square-shaped area equivalent to A_2 and neglecting the remaining portion.

(D) This answer incorrectly assumes that the increase in vertical stress at point A is the same as the increase in vertical pressure at the surface ($p = 4000$ lbf/ft^2).

10 Deep Foundations

GROUP EFFECTS

PROBLEM 1

Two 36 in diameter piles were designed for an allowable capacity on an individual basis, but were to be spaced 72 in on center. However, the two piles were mistakenly constructed at 60 in on center. Given that the structural loading conditions, length of embedment, and soil conditions are the same for both piles, determine how the two piles will act.

(A) They will act in a group capacity.

(B) They will act in an individual capacity.

(C) It cannot be determined, due to lack of soil profile information.

(D) It cannot be determined, due to lack of pile dimensional information.

Hint: Compare the circumference of the two piles as a group to the two piles acting individually.

PROBLEM 2

Twelve 24 in diameter round piles are to be driven vertically below the bottom of the river to a depth of 50 ft for support of a bridge abutment. The piles are to be spaced 6 ft on center in a 3 by 4 pattern in the soil profile shown.

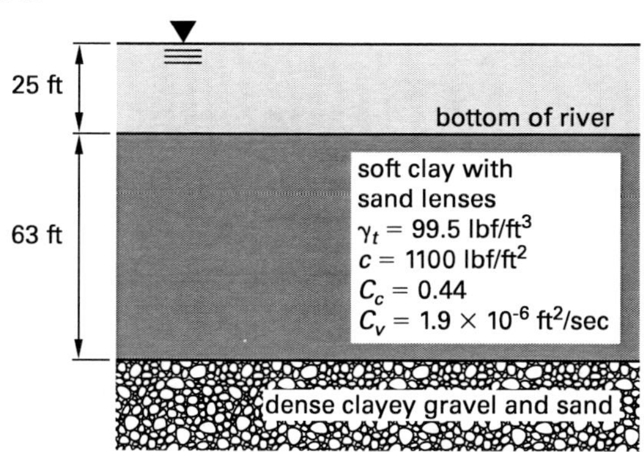

Using the average α-method to calculate the friction capacity and using a factor of safety equal to 3, approximate the allowable capacity of the group.

(A) 180 kips

(B) 1300 kips

(C) 2800 kips

(D) 6500 kips

Hint: Because only one layer of soil with relevant engineering properties exists, the undrained shear strength is the same throughout the entire driven length of the pile.

PROBLEM 3

A 2 by 2 pile group extends to a depth of 24 ft in the soil profile shown. The group load is 120 kips or 30 kips per pile. The individual 18 in diameter piles are spaced 24 in on center.

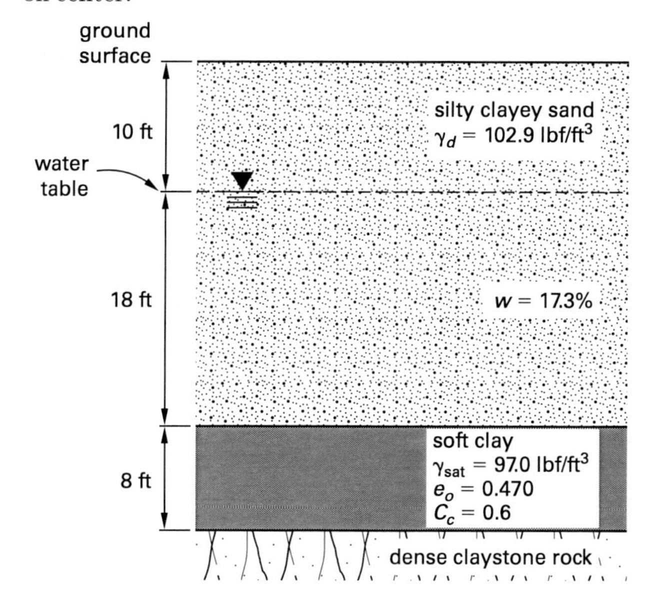

Knowing that group action is occurring, determine the approximate settlement of the piles if the water table elevation remains unchanged.

(A) 0.2 in

(B) 1 in

(C) 2 in

(D) 5 in

Hint: The 60° method is typically used for evaluating the pressure distribution at depth for group action.

PROBLEM 4

A group of 30 in diameter caissons is constructed for support of a bridge. The group is arranged in a 3 by 4 pattern and extends to a depth of 30 ft below the reservoir bottom in the soil profile shown. The structural load on the group is 300 kips, and the individual drilled piers are spaced 36 in on center.

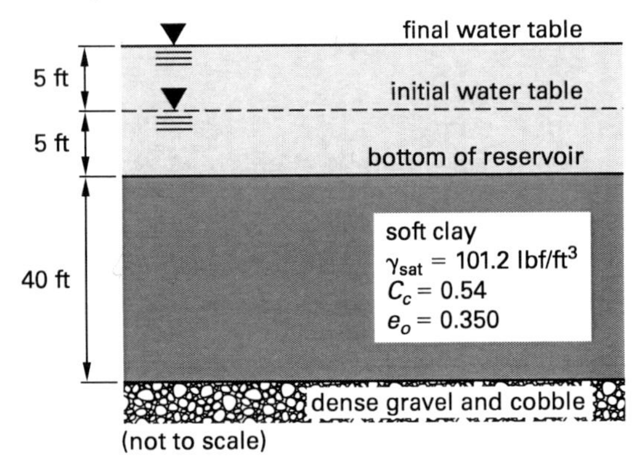

Under group action, calculate the approximate settlement of the group if the water level is raised 5 ft after installation during structural loading of the bridge.

(A) 4 in

(B) 8 in

(C) 10 in

(D) 19 in

Hint: Use a common pressure distribution method for evaluating the consolidation pressure at depth for group action.

INSTALLATION METHODS/HAMMER SELECTION

PROBLEM 5

A double-acting hammer was able to drive an 8 in diameter closed-tip steel pile to a depth of 40 ft at a rate of ½ in per blow in the soil profile shown.

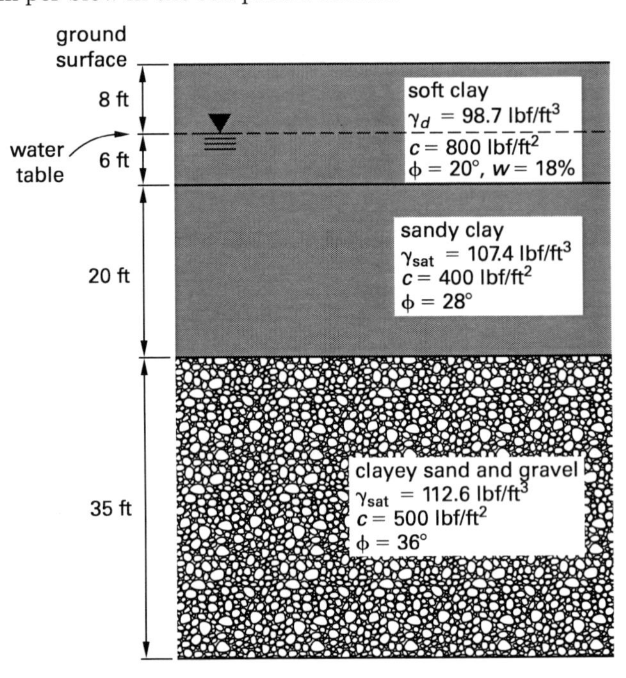

Given a striking weight of 12.5 kips with a hammer-energy rating of 18 ft-kips, approximate the allowable pile capacity assuming the driving cap and block each weigh 1.5 kips.

(A) 30 kips

(B) 60 kips

(C) 250 kips

(D) 720 kips

Hint: The dynamic driving data can be used to calculate the capacity.

LATERAL LOAD AND DEFORMATIONAL ANALYSIS

PROBLEM 6

Approximate the maximum allowable soil reaction (per unit length of pier) for a 5 ft diameter cast-in-place drilled pier if the modulus of subgrade reaction for the soil at the point of maximum lateral deflection is 120 lbf/in^3.

(A) 0.50 lbf/in

(B) 24 lbf/in

(C) 600 lbf/in

(D) 7200 lbf/in

Hint: The diameter of the pier is the horizontal component of the contact area at the point of lateral deflection.

PILE AND DRILLED-SHAFT LOAD TESTING

PROBLEM 7

The graphical data shown was obtained from laboratory testing and developed based on research.

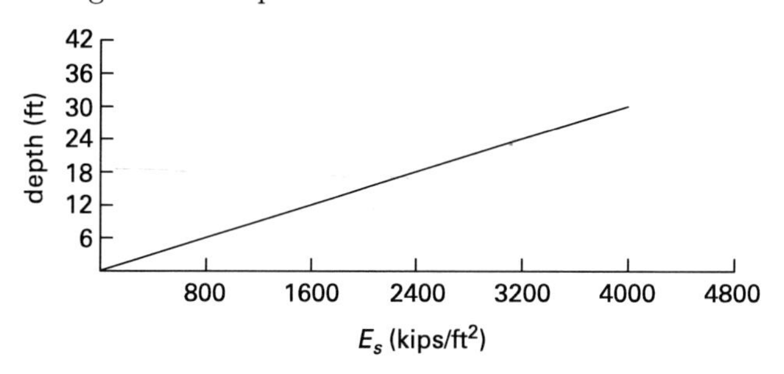

Young's modulus vs. depth

Approximate the soil reaction for a unit lateral deflection of a 14 in square precast concrete pile at a depth of 18 ft.

(A) 300 lbf/in

(B) 1100 lbf/in

(C) 1900 lbf/in

(D) 17,000 lbf/in

Hint: The modulus of subgrade reaction is determined from the slope of the graph.

PROBLEM 8

A p-y curve is generated from a testing program as shown for a driven pile at a depth of 5.5 ft.

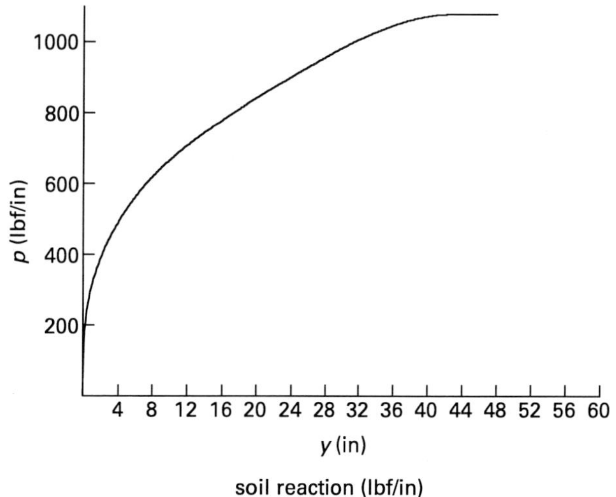

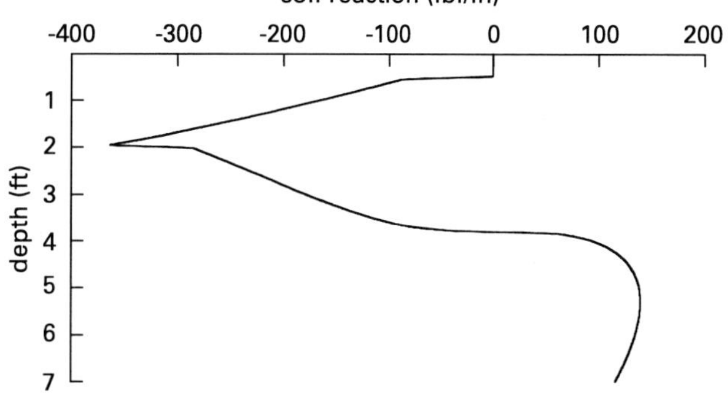

Use the soil reaction versus depth curve to determine the percentage of maximum soil reaction (p/p_{max}) for the pile at 5.5 ft below the ground surface.

(A) 0.13%

(B) 7.5%

(C) 15%

(D) 63%

Hint: The maximum allowable soil reaction value corresponds to the maximum lateral deflection.

PROBLEM 9

A jurisdictional building code requires that maximum allowable pile loads are to be determined based on pile load testing data. The code specifies that the maximum allowable pile load is to be one-third the value of the test load that results in a total settlement value of not more than 0.002 in/kip.

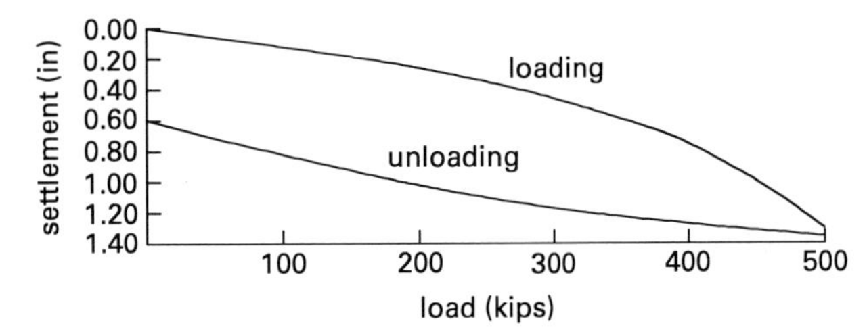

Given the pile load test data shown, use the code criteria to approximate the maximum allowable pile load.

(A) 150 kips

(B) 230 kips

(C) 450 kips

(D) 500 kips

Hint: Use the maximum settlement criteria with the total load achieved during the pile load testing to determine the limiting settlement value and corresponding load.

SINGLE ELEMENT AXIAL CAPACITY

PROBLEM 10

An 18 in diameter round steel pile is to be driven to a depth of 30 ft in the soil profile shown. For clean steel, $c_A = 0.7c$. For driven piles, assume the coefficient of lateral earth pressure is equal to 1.

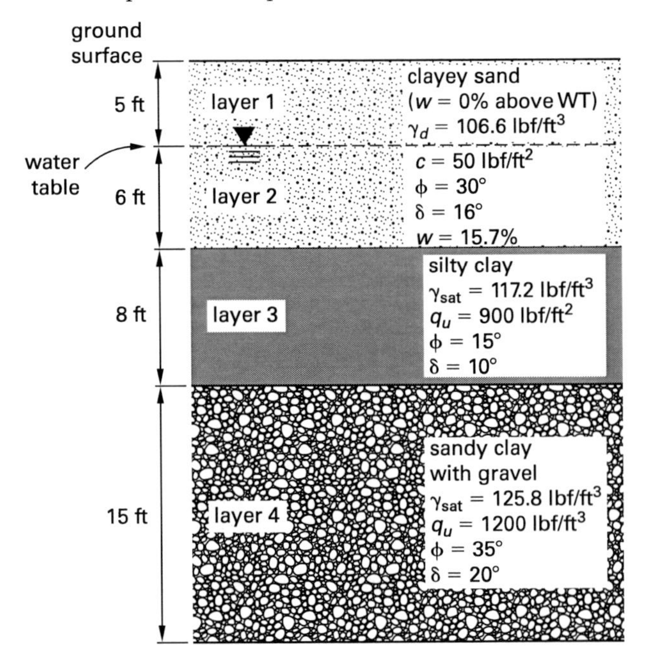

Using the Mohr-Coulomb equation for failure, approximate the skin friction capacity of the pile.

(A) 70 kips

(B) 82 kips

(C) 110 kips

(D) 120 kips

Hint: Treat the soil-structure interaction of the surface of the pile similarly to the failure plane in a direct shear test by finding the shear strength of the soil along that plane.

PROBLEM 11

A group of 12 in diameter precast concrete piles is to be driven to a depth of 140 ft below the water level for support of a wharf structure. The piles are to be spaced approximately 8 ft apart in the soil profile shown.

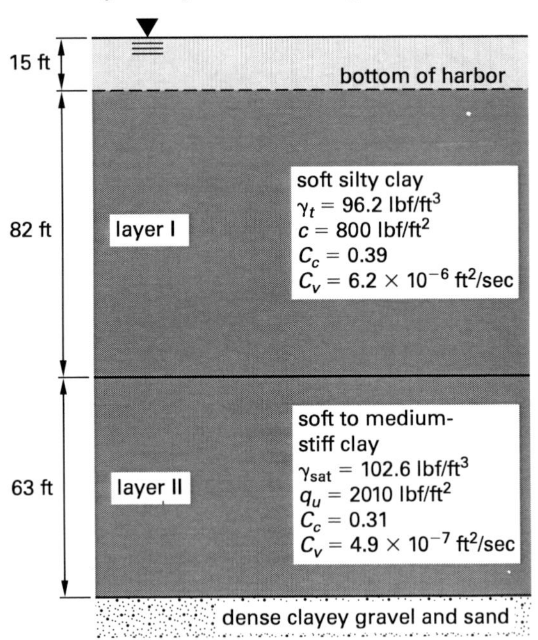

Using typical Meyerhof values from the β-method and a factor of safety equal to 3, approximate the allowable capacity of a single pile.

(A) 50 kips

(B) 55 kips

(C) 82 kips

(D) 150 kips

Hint: Determine the friction capacity for each layer and the tip capacity.

PROBLEM 12

A group of 30 in diameter drilled piers arranged 2 by 2 and 3 ft on center is to be installed to a depth of 25 ft in the soil profile shown.

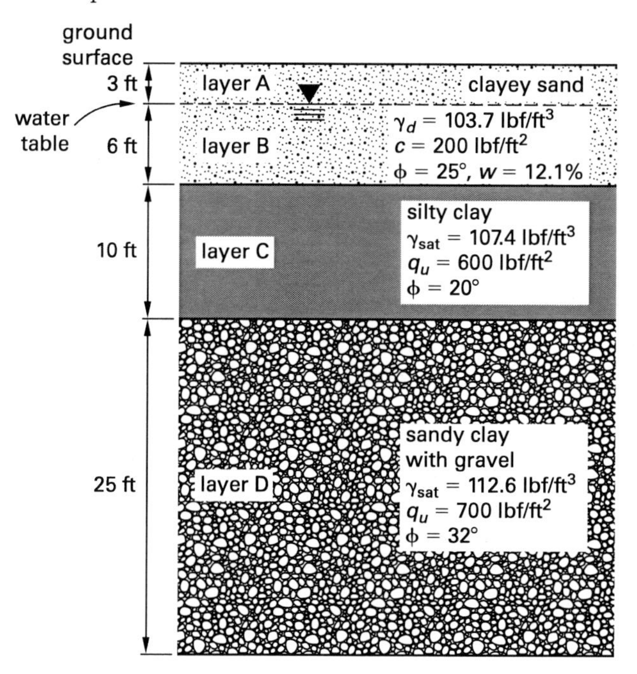

Approximate the ultimate capacity of an individual drilled pier in the group assuming a factor of safety of 2 and block behavior with uniform concentric structural loading. Use typical Meyerhof values from the β-method to calculate the friction capacity where necessary.

(A) 23 kips

(B) 59 kips

(C) 62 kips

(D) 190 kips

Hint: Determine the average cohesion for each layer.

PROBLEM 13

A 14 in square precast concrete pile is to be driven to a depth of 45 ft in the soil profile shown.

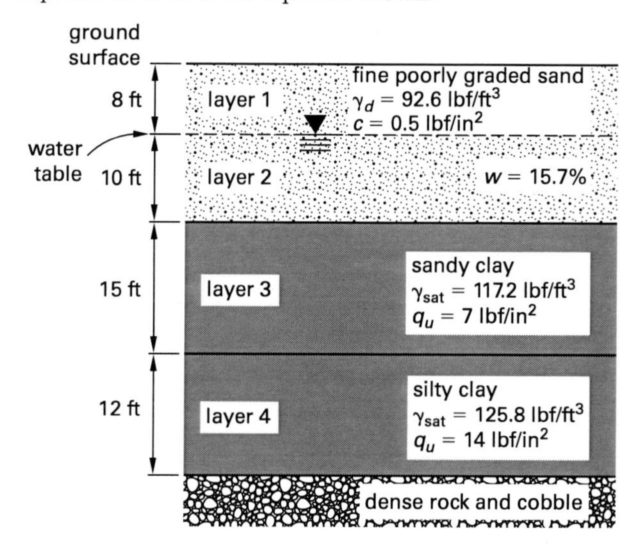

Using average values from the α-method, approximate the total capacity of the pile.

(A) 20 kips

(B) 80 kips

(C) 100 kips

(D) 160 kips

Hint: Determine the friction capacity for each layer and the tip capacity.

SOLUTION 1

If the circumference of the two piles as a group is greater than the circumference of the two piles individually, the piles will retain 100% efficiency and no group reduction is required.

Calculate the group perimeter.

$$\begin{aligned} C_{\text{group}} &= \pi d + 2(\text{spacing}) \\ &= \pi(36\ \text{in}) + (2)(60\ \text{in}) \\ &= 233\ \text{in} \end{aligned}$$

Calculate the circumferential perimeter.

$$\begin{aligned} C_{\text{ind-pile}\times 2} &= 2\pi d = 2\pi(36\ \text{in}) \\ &= 226\ \text{in} \end{aligned}$$

The group perimeter is greater than the sum of the two individual circumferences. Therefore, the two piles will act individually, as designed.

The answer is (B).

Why Other Options Are Wrong

(A) This option mistakenly calculates the group perimeter by omitting the perimeter distance between the two piles and determining that the group perimeter is less than the sum of the two individual circumferences.

(C) This option is incorrect because soil profile information is not required, since the perimeter and circumference information is constant for each evaluation. Therefore, this determination can be made from a comparison of the required circumference length that would need to be mobilized to activate needed shear strength and achieve the proper friction capacity for the same loading conditions.

(D) This option is incorrect because the pile lengths are the same and because pile dimension information is not required, since the perimeter and circumference information is constant for each evaluation.

SOLUTION 2

The pile penetrates below the river bottom a depth of 50 ft, or 75 ft below the water surface. The soil is cohesive; therefore, group capacity is taken as the smaller value of the sum of the individual capacities or the total (ultimate) group pile capacity, assuming block action. The ultimate block action capacity is the sum of the friction (side) capacity and the end bearing (tip) capacity.

$$Q_{\text{ult}} = Q_s + Q_p$$

The allowable capacity is the ultimate capacity reduced by a factor of safety.

$$Q_a = \frac{Q_{\text{ult}}}{F} = \frac{Q_s + Q_p}{3}$$

Use the α-method as specified for determination of the friction capacity for an individual pile.

$$\begin{aligned} \alpha &= 0.83 \\ f_s &= \alpha c \\ Q_{s,\text{pile}} &= f_s \pi B L = \alpha c \pi B L \\ &= (0.83)\left(1100\ \frac{\text{lbf}}{\text{ft}^2}\right)\pi(24\ \text{in})(50\ \text{ft})\left(\frac{1\ \text{ft}}{12\ \text{in}}\right) \\ &= 286{,}827\ \text{lbf} \quad (287\ \text{kips}) \end{aligned}$$

The end bearing value for an individual pile can be calculated. N_c is taken as 9 for driven piles.

$$\begin{aligned} Q_{p,\text{pile}} &= N_c c A_p \\ &= (9)\left(1100\ \frac{\text{lbf}}{\text{ft}^2}\right)\left(\frac{\pi}{4}\right)(24\ \text{in})^2\left(\frac{1\ \text{ft}}{12\ \text{in}}\right)^2 \\ &= 31{,}102\ \text{lbf} \quad (31\ \text{kips}) \end{aligned}$$

The individual pile ultimate capacity can be calculated.

$$\begin{aligned} Q_{\text{ult,pile}} &= Q_s + Q_p = 287\ \text{kips} + 31\ \text{kips} \\ &= 318\ \text{kips} \end{aligned}$$

The allowable capacity for an individual pile is the ultimate capacity reduced by a factor of safety.

$$\begin{aligned} Q_{a,\text{pile}} &= \frac{Q_{\text{ult}}}{F} = \frac{318\ \text{kips}}{3} \\ &= 106\ \text{kips} \end{aligned}$$

The quantity of piles in the group is 12. The total allowable capacity can be calculated.

$$\begin{aligned} Q_{a,\text{group}} &= nQ_{a,\text{pile}} = (12)(106\ \text{kips}) \\ &= 1272\ \text{kips} \quad (1270\ \text{kips}) \end{aligned}$$

The friction capacity for block action can be calculated.

$$\begin{aligned} Q_{s,\text{block}} &= 2(b + w)Lc \\ &= (2)(14\ \text{ft} + 20\ \text{ft})(50\ \text{ft})\left(1100\ \frac{\text{lbf}}{\text{ft}^2}\right) \\ &= 3.74 \times 10^6\ \text{lbf} \quad (3740\ \text{kips}) \end{aligned}$$

The end bearing capacity for block action can be calculated.

$$\begin{aligned} Q_{p,\text{block}} &= N_c c b w \\ &= (9)\left(1100\ \frac{\text{lbf}}{\text{ft}^2}\right)(14\ \text{ft})(20\ \text{ft}) \\ &= 2{,}772{,}000\ \text{lbf} \quad (2772\ \text{kips}) \end{aligned}$$

The total block action capacity can be determined.

$$\begin{aligned} Q_{\text{ult,block}} &= Q_{s,\text{block}} + Q_{p,\text{block}} \\ &= 3740\ \text{kips} + 2772\ \text{kips} \\ &= 6512\ \text{kips} \end{aligned}$$

$$\begin{aligned} Q_{a,\text{block}} &= \frac{Q_{\text{ult,block}}}{3} \\ &= \frac{6512\ \text{kips}}{3} \\ &= 2171\ \text{kips} \quad (2170\ \text{kips}) \end{aligned}$$

The value for allowable capacity determined from block action is compared to the value obtained from individual pile capacity, and the smaller value is chosen.

$$\begin{aligned} Q_{a,\text{block}} &= 2170\ \text{kips} \\ Q_{a,\text{group}} &= 1270\ \text{kips} \quad (1300\ \text{kips}) \\ Q_{a,\text{block}} &> Q_{a,\text{group}} \end{aligned}$$

The answer is (B).

Why Other Options Are Wrong

(A) This option mistakenly divides the allowable block action capacity by 12 to determine the individual pile capacity instead of the group capacity.

(C) This option mistakenly includes the water depth in the embedment value for calculation of the ultimate group capacity value.

(D) This option fails to reduce the ultimate block action capacity by the factor of safety.

SOLUTION 3

The equation for consolidation settlement can be expressed as

$$S = C_c\left(\frac{H_o}{1+e_o}\right)\log_{10}\frac{p_o' + \Delta p_v'}{p_o'}$$

Calculate the initial effective stress at the midpoint of the clay layer.

$$\begin{aligned} p_o' &= p_{\text{clay,midpoint}} - u_o \\ &= \gamma_{d,\text{sand}} H_{\text{sand,dry}} \\ &\quad + \left(\gamma_{d,\text{sand}}(1+w) H_{\text{sand,wet}}\right) \\ &\quad + \gamma_{\text{sat,clay}} H_{\text{clay,midpoint}} - \gamma_w H_w \\ &= \left(102.9\ \frac{\text{lbf}}{\text{ft}^3}\right)(10\ \text{ft}) + \left(102.9\ \frac{\text{lbf}}{\text{ft}^3}\right) \\ &\quad \times (1+0.173)(18\ \text{ft}) + \left(97\ \frac{\text{lbf}}{\text{ft}^3}\right)\left(\frac{8\ \text{ft}}{2}\right) \\ &\quad - \left(62.4\ \frac{\text{lbf}}{\text{ft}^3}\right)\left(18\ \text{ft} + \frac{8\ \text{ft}}{2}\right) \\ &= 2217\ \text{lbf/ft}^2 \end{aligned}$$

The 60° method should be used to evaluate the pressure distribution from a point $\frac{2}{3}L_{\text{pile group}}$ below the top of the pile group for purposes of calculating the consolidation in the clay layer. Find the area of influence at the midpoint of the clay layer (area will be 16 ft below the depth of $\frac{2}{3}L_{\text{pile group}}$).

$$\begin{aligned} A &= \left(B + 2(h\cot 60°)\right)\left(L + 2(h\cot 60°)\right) \\ &= \left(3.5\ \text{ft} + (2)(16\ \text{ft})\cot 60°\right) \\ &\quad \times \left(3.5\ \text{ft} + (2)(16\ \text{ft})\cot 60°\right) \\ &= 482.9\ \text{ft}^2 \end{aligned}$$

Determine the change in effective stress at the midpoint of the clay layer due to the total increase in load from the pile group.

$$\begin{aligned} \Delta p_v' &= \frac{P_{\text{pile group}}}{A} = \frac{120\ \text{kips}}{482.9\ \text{ft}^2} \\ &= 0.2485\ \text{kips/ft}^2 \quad (248.5\ \text{lbf/ft}^2) \end{aligned}$$

Determine the total expected settlement.

$$\begin{aligned} S &= C_c\left(\frac{H_o}{1+e_o}\right)\log_{10}\frac{p_o' + \Delta p_v'}{p_o'} \\ &= (0.6)\left(\frac{8\ \text{ft}}{1+0.470}\right) \\ &\quad \times \log_{10}\frac{2217\ \frac{\text{lbf}}{\text{ft}^2} + 248.5\ \frac{\text{lbf}}{\text{ft}^2}}{2217\ \frac{\text{lbf}}{\text{ft}^2}}\left(12\ \frac{\text{in}}{\text{ft}}\right) \\ &= 1.8\ \text{in} \quad (2\ \text{in}) \end{aligned}$$

The answer is (C).

Why Other Options Are Wrong

(A) This solution is incorrect because it is the approximate settlement value, in feet, obtained by erroneously failing to convert the calculation to inches. The units are not correct.

(B) This solution erroneously uses half of the clay layer thickness for determining the settlement of the clay layer.

(D) This incorrect solution determines the pressure distribution from the bottom of the pile group rather than from $\frac{2}{3}L_{\text{pile group}}$.

SOLUTION 4

The equation for consolidation settlement can be expressed as

$$S = C_c\left(\frac{H_o}{1+e_o}\right)\log_{10}\frac{p_o' + \Delta p_v'}{p_o'}$$

Calculate the initial effective stress at the midpoint of the portion of the clay layer that will consolidate between the base of the drilled pier group and the gravel and cobble layer.

$$\begin{aligned} p_o' &= p_{\text{clay,midpoint}} - u_o \\ &= \gamma_{\text{sat,clay}}H_{\text{clay,midpoint}} + (\gamma_w)(D_{\text{res-initial}}) - \gamma_w H_w \\ &= \left(101.2\ \frac{\text{lbf}}{\text{ft}^3}\right)(30\text{ ft}) + \left(62.4\ \frac{\text{lbf}}{\text{ft}^3}\right)(5\text{ ft}) \\ &\quad - \left(62.4\ \frac{\text{lbf}}{\text{ft}^3}\right)(35\text{ ft}) \\ &= 1164\ \text{lbf/ft}^2 \end{aligned}$$

A common pressure distribution method, the 60° method, can be used to evaluate the pressure distribution from a point $\frac{2}{3}L_{\text{pile group}}$ below the top of the pier group for purposes of calculating the consolidation in the clay layer. Find the area of influence at the midpoint of the clay layer (10 ft below the depth of $\frac{2}{3}L_{\text{pile group}}$).

$$\begin{aligned} A &= \big(B + 2(h\cot 60°)\big)\big(L + 2(h\cot 60°)\big) \\ &= \left((102\text{ in})\left(\frac{1\text{ ft}}{12\text{ in}}\right) + (2)(10\text{ ft})\cot 60°\right) \\ &\quad \times\left((138\text{ in})\left(\frac{1\text{ ft}}{12\text{ in}}\right) + (2)(10\text{ ft})\cot 60°\right) \\ &= 462\text{ ft}^2 \end{aligned}$$

Determine the change in effective stress at the midpoint of the clay layer due to the total increase in load from the pier group.

$$\begin{aligned} \Delta p'_{v,\text{pier load}} &= \frac{P_{\text{pier load}}}{A} = \frac{300\text{ kips}}{462\text{ ft}^2} \\ &= 0.649\text{ kips/ft}^2 \quad (650\text{ lbf/ft}^2) \end{aligned}$$

The water level is to be simultaneously raised as the structural loading is applied after the drilled piers are installed. Determine the additional change in effective stress due to the water level change.

$$\begin{aligned} \Delta p'_{v,\text{new water height}} &= (p_{\text{clay,midpoint}} - u_{\text{new}}) - p_o' \\ &= \gamma_{\text{sat,clay}}H_{\text{clay,midpoint}} + (\gamma_w)(D_{\text{res-final}}) \\ &\quad - \gamma_w H_{w,\text{new}} - p_o' \\ &= \left(101.2\ \frac{\text{lbf}}{\text{ft}^3}\right)(30\text{ ft}) \\ &\quad + \left(62.4\ \frac{\text{lbf}}{\text{ft}^3}\right)(10\text{ ft}) \\ &\quad - \left(62.4\ \frac{\text{lbf}}{\text{ft}^3}\right)(40\text{ ft}) - 1164\ \frac{\text{lbf}}{\text{ft}^2} \\ &= 0\ \text{lbf/ft}^2 \end{aligned}$$

A negligible change in effective stress occurs due to the increase in water level. The total change in effective stress can be calculated.

$$\begin{aligned} \Delta p_v' &= \Delta p'_{v,\text{pier load}} + \Delta p'_{v,\text{new water height}} \\ &= 650\ \frac{\text{lbf}}{\text{ft}^2} + 0\ \frac{\text{lbf}}{\text{ft}^2} \\ &= 650\ \text{lbf/ft}^2 \end{aligned}$$

Determine the total expected settlement.

$$\begin{aligned} S &= C_c\left(\frac{H_o}{1+e_o}\right)\log_{10}\frac{p_o' + \Delta p_v'}{p_o'} \\ &= (0.54)\left(\frac{20\text{ ft}}{1 + 0.350}\right) \\ &\quad \times \log_{10}\left(\frac{1164\ \frac{\text{lbf}}{\text{ft}^2} + 650\ \frac{\text{lbf}}{\text{ft}^2}}{1164\ \frac{\text{lbf}}{\text{ft}^2}}\right)\left(12\ \frac{\text{in}}{\text{ft}}\right) \\ &= 18.5\text{ in} \quad (19\text{ in}) \end{aligned}$$

The answer is (D).

Why Other Options Are Wrong

(A) This solution erroneously uses half the clay layer thickness for determining the settlement of the clay layer.

(B) This solution mistakenly calculates the pressure distribution at 5 ft above the bottom of the clay layer rather than at the midpoint of the clay layer (between the location of $\frac{2}{3}L_{\text{pile group}}$ and the hard layer).

(C) This solution erroneously adds the change in effective stress due to the increased water level rather than subtracting it (algebraically).

SOLUTION 5

No wall thickness is given for the steel pile. To show that the driven weight (pile weight) is less than the striking weight (hammer weight), assume the pile is solid steel and use a reasonable value for the unit weight of steel, 490 lbf/ft^3.

$$\begin{aligned} W_P &= V_P\gamma_{\text{steel}} + \text{cap} + \text{capblock} \\ &= A_P L_P \gamma_{\text{steel}} + \text{cap} + \text{capblock} \end{aligned}$$

$$\begin{aligned} A_p &= \frac{\pi}{4}D^2 = \left(\frac{\pi}{4}\right)(8\ \text{in})^2\left(\frac{1\ \text{ft}}{12\ \text{in}}\right)^2 \\ &= 0.35\ \text{ft}^2 \end{aligned}$$

$$\begin{aligned} W_P &= A_P L_P \gamma_{\text{steel}} + \text{cap} + \text{capblock} \\ &= (0.35\ \text{ft}^2)(40\ \text{ft})\left(490\ \frac{\text{lbf}}{\text{ft}^3}\right)\left(\frac{1\ \text{kip}}{1000\ \text{lbf}}\right) \\ &\quad + 1.5\ \text{kips} + 1.5\ \text{kips} \\ &= 9.86\ \text{kips} \quad (9.9\ \text{kips}) \end{aligned}$$

Use the *Engineering News Record* (ENR) driving equation to calculate the allowable capacity, given the average distance per blow, S (required to be in inches), and the hammer energy.

$$\begin{aligned} Q_a &= \frac{2E}{S+0.1} = \frac{(2)(18\ \text{ft-kips})}{0.5\ \text{in} + 0.1} \\ &= 60\ \text{kips} \end{aligned}$$

The answer is (B).

Why Other Options Are Wrong

(A) This option mistakenly interchanges the hammer energy and hammer weight values.

(C) This option mistakenly converts the distance per blow value to feet and calculates the wrong allowable pile capacity.

(D) This option mistakenly converts the hammer energy value to kip-inches and calculates the wrong allowable pile capacity.

SOLUTION 6

The maximum allowable soil reaction per inch of pier length is the modulus of subgrade reaction multiplied by the pier diameter.

$$\begin{aligned} p_{\max} &= KD\Delta \\ &= \left(120\ \frac{\text{lbf}}{\text{in}^3}\right)(5\ \text{ft})\left(12\ \frac{\text{in}}{\text{ft}}\right)(1\ \text{in}) \\ &= 7200\ \text{lbf/in per inch of pier length} \end{aligned}$$

The answer is (D).

Why Other Options Are Wrong

(A) This option mistakenly divides the diameter by the modulus of subgrade reaction. The units do not work out correctly.

(B) This option mistakenly divides the modulus of subgrade reaction by the diameter and does not convert to inches. The units do not work out correctly.

(C) This option fails to convert the answer to inches. The units do not work out correctly.

SOLUTION 7

From the graphical data presented, observe that Young's modulus varies linearly with depth. Therefore, determine the modulus of subgrade reaction for the soil by finding the slope of the line.

$$\begin{aligned} K &= \frac{\Delta E_s}{\Delta D} \\ &= \frac{3200\ \frac{\text{kips}}{\text{ft}^2} - 800\ \frac{\text{kips}}{\text{ft}^2}}{24\ \text{ft} - 6\ \text{ft}} \\ &= 133\ \text{kips/ft}^3 \end{aligned}$$

Due to the linear assumption of Young's modulus, the modulus of subgrade reaction, assumed to be as plotted, is the same throughout the pile length and is independent of depth. The soil reaction corresponding to a unit lateral deflection can be determined by finding the soil reaction corresponding to a 1 in lateral deflection of the pile. The soil reaction is calculated from the modulus of subgrade reaction multiplied by the pile width and the lateral deflection of 1 in.

$$\begin{aligned} p = Kd\Delta &= \left(133\ \frac{\text{kips}}{\text{ft}^3}\right)\left(\frac{1\ \text{ft}}{12\ \text{in}}\right)^3(14\ \text{in})(1\ \text{in}) \\ &= 1.078\ \text{kips/in} \quad (1100\ \text{lbf/in}) \end{aligned}$$

The answer is (B).

Why Other Options Are Wrong

(A) This option is incorrect because it does not determine the modulus of subgrade reaction from the graphical data. Instead, it multiplies the value for Young's modulus at 18 ft by the depth of 18 ft. The units do not work out correctly.

(C) This option fails to convert the answer to inches and pounds-force. The units do not work out correctly.

(D) This option is incorrect because it multiplies the value for Young's modulus at a depth of 18 ft by a unit value of lateral deflection (1 in) without converting to pounds-force.

SOLUTION 8

The maximum allowable soil reaction occurs at the maximum lateral deflection of the pile. From the *p-y* curve for a depth of 5.5 ft, the maximum allowable soil reaction per inch of pier length is estimated to be approximately 1000 lbf/in as shown.

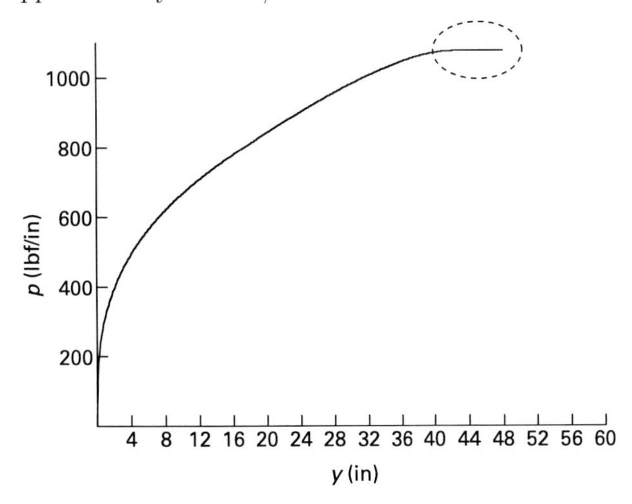

The actual soil reaction for the pile at 5.5 ft is determined to be approximately 150 lbf/in as shown.

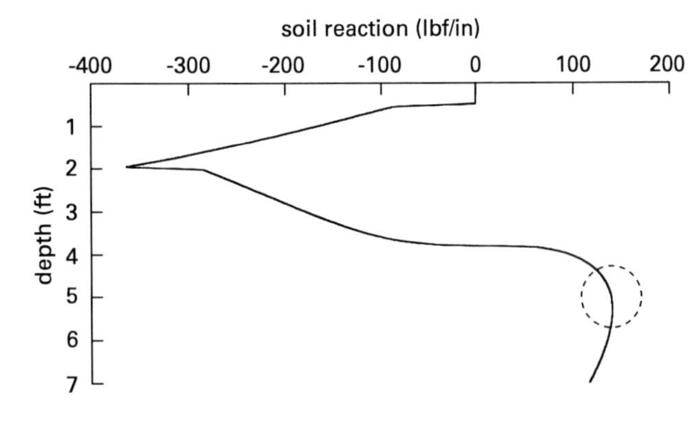

The percentage of the maximum soil reaction can be calculated.

$$\frac{p}{p_{\max}} \times 100\% = \frac{150\ \frac{\text{lbf}}{\text{in}}}{1000\ \frac{\text{lbf}}{\text{in}}} \times 100\% = 15\%$$

The answer is (C).

Why Other Options Are Wrong

(A) This option neglects to express the result as a percentage.

(B) This option mistakenly inverts the calculation and neglects to express the result as a percentage.

(D) This option mistakenly converts the actual soil reaction to a per foot value rather than a per inch value when inverting the calculation.

SOLUTION 9

The maximum settlement value is calculated using the code settlement criteria.

$$\begin{aligned} S_{\max} &= P_{\text{test load,max}}\left(0.002\ \frac{\text{in}}{\text{kip}}\right) \\ &= (500\ \text{kips})\left(0.002\ \frac{\text{in}}{\text{kip}}\right) \\ &= 1\ \text{in} \end{aligned}$$

Using the graphed data, the test load corresponding to the maximum settlement value of 1 in is determined.

$$P_{\text{test load,set}} = 450\ \text{kips}$$

The maximum allowable pile load can be calculated using the code criteria.

$$\begin{aligned} P_a &= \tfrac{1}{3}P_{\text{test load,set}} \\ &= \left(\frac{1}{3}\right)(450\ \text{kips}) \\ &= 150\ \text{kips} \end{aligned}$$

The answer is (A).

Why Other Options Are Wrong

(B) This option mistakenly takes half the corresponding test load instead of one third.

(C) This option fails to reduce the corresponding test load by the required one third.

(D) This option mistakenly assumes the corresponding test load is the maximum test load.

SOLUTION 10

In the case of this friction capacity problem, the failure plane can be considered to be the surface area of the pile. The Mohr-Coulomb equation determines the shear strength of the soil in response to normal stress along the failure plane and can be expressed with the coefficient of lateral earth pressure to calculate the skin friction.

$$f_s = c_A + k(\sigma_v \tan\delta)$$

The friction capacity for an individual layer can be expressed using the equation

$$Q_s = f_s A_s = (c_A + k\sigma_v \tan\delta)\pi DL$$

Determine the overburden (confining) pressure at the midpoint of each soil layer, using effective stress analysis where necessary.

$$\begin{aligned}\sigma_{v1,\text{ave}} &= \gamma_{d,1}\left(\frac{L_1}{2}\right) = \left(106.6\ \frac{\text{lbf}}{\text{ft}^3}\right)\left(\frac{5\ \text{ft}}{2}\right)\\ &= 266.5\ \text{lbf/ft}^2\end{aligned}$$

$$\begin{aligned}\sigma'_{v2,\text{ave}} &= \gamma_{d,1}L_1 + \left(\gamma_{d,1}(1+w_2) - \gamma_w\right)\left(\frac{L_2}{2}\right)\\ &= \left(106.6\ \frac{\text{lbf}}{\text{ft}^3}\right)(5\ \text{ft})\\ &\quad + \left(\begin{array}{c}\left(106.6\ \frac{\text{lbf}}{\text{ft}^3}\right)(1+0.157)\\ -62.4\ \frac{\text{lbf}}{\text{ft}^3}\end{array}\right)\left(\frac{6\ \text{ft}}{2}\right)\\ &= 715.8\ \text{lbf/ft}^2\end{aligned}$$

$$\begin{aligned}\sigma'_{v3,\text{ave}} &= \gamma_{d,1}L_1 + \left(\gamma_{d,1}(1+w_2) - \gamma_w\right)L_2\\ &\quad + (\gamma_{\text{sat},3} - \gamma_w)\left(\frac{L_3}{2}\right)\\ &= \left(106.6\ \frac{\text{lbf}}{\text{ft}^3}\right)(5\ \text{ft})\\ &\quad + \left(\left(106.6\ \frac{\text{lbf}}{\text{ft}^3}\right)(1+0.157) - 62.4\ \frac{\text{lbf}}{\text{ft}^3}\right)\\ &\quad \times(6\ \text{ft}) + \left(117.2\ \frac{\text{lbf}}{\text{ft}^3} - 62.4\ \frac{\text{lbf}}{\text{ft}^3}\right)\left(\frac{8\ \text{ft}}{2}\right)\\ &= 1117.8\ \text{lbf/ft}^2\end{aligned}$$

$$\begin{aligned}\sigma'_{v4,\text{ave}} &= \gamma_d L_1 + \left(\gamma_d(1+w) - \gamma_w\right)L_2\\ &\quad + (\gamma_{\text{sat},3} - \gamma_w)L_3 + (\gamma_{\text{sat},4} - \gamma_w)\left(\frac{L_4}{2}\right)\\ &= \left(106.6\ \frac{\text{lbf}}{\text{ft}^3}\right)(5\ \text{ft})\\ &\quad + \left(\left(106.6\ \frac{\text{lbf}}{\text{ft}^3}\right)(1+0.157) - 62.4\ \frac{\text{lbf}}{\text{ft}^3}\right)\\ &\quad \times(6\ \text{ft}) + \left(117.2\ \frac{\text{lbf}}{\text{ft}^3} - 62.4\ \frac{\text{lbf}}{\text{ft}^3}\right)(8\ \text{ft})\\ &\quad + \left(125.8\ \frac{\text{lbf}}{\text{ft}^3} - 62.4\ \frac{\text{lbf}}{\text{ft}^3}\right)\left(\frac{11\ \text{ft}}{2}\right)\\ &= 1685.7\ \text{lbf/ft}^2\end{aligned}$$

The coefficient of lateral earth pressure assumed for the driven pile is 1. Determine the friction capacity for each soil layer. The cohesions for both layers 1 and 2 are equivalent; however, the unconfined compressive strength is presented for layers 2 and 3.

$$\begin{aligned}f_{s1} &= c_{A1} + k_1(\sigma_{v1,\text{ave}}\tan\delta_1)\\ &= (0.7)\left(50\ \frac{\text{lbf}}{\text{ft}^2}\right) + (1)\left(266.5\ \frac{\text{lbf}}{\text{ft}^2}\right)\tan 16^\circ\\ &= 111.4\ \text{lbf/ft}^2\end{aligned}$$

$$\begin{aligned}f_{s2} &= c_{A2} + k_2(\sigma'_{v2,\text{ave}}\tan\delta_2)\\ &= (0.7)\left(50\ \frac{\text{lbf}}{\text{ft}^2}\right) + (1)\left(715.8\ \frac{\text{lbf}}{\text{ft}^2}\right)\tan 16^\circ\\ &= 240.3\ \text{lbf/ft}^2\end{aligned}$$

$$\begin{aligned}f_{s3} &= c_{A3} + k_3(\sigma'_{v3,\text{ave}}\tan\delta_3)\\ &= (0.7)\left(450\ \frac{\text{lbf}}{\text{ft}^2}\right) + (1)\left(1117.8\ \frac{\text{lbf}}{\text{ft}^2}\right)\tan 10^\circ\\ &= 512.1\ \text{lbf/ft}^2\end{aligned}$$

$$\begin{aligned}f_{s4} &= c_{A4} + k_4(\sigma'_{v4,\text{ave}}\tan\delta_4)\\ &= (0.7)\left(600\ \frac{\text{lbf}}{\text{ft}^2}\right) + (1)\left(1685.7\ \frac{\text{lbf}}{\text{ft}^2}\right)\tan 20^\circ\\ &= 1033.5\ \text{lbf/ft}^2\end{aligned}$$

Determine the pile surface area for each soil layer.

$$\begin{aligned}A_{s1} &= \pi DL_1 = \pi(18\ \text{in})(5\ \text{ft})\left(\frac{1\ \text{ft}}{12\ \text{in}}\right)\\ &= 23.6\ \text{ft}^2\end{aligned}$$

$$A_{s2} = \pi DL_2 = \pi(18 \text{ in})(6 \text{ ft})\left(\frac{1 \text{ ft}}{12 \text{ in}}\right) = 28.3 \text{ ft}^2$$

$$A_{s3} = \pi DL_3 = \pi(18 \text{ in})(8 \text{ ft})\left(\frac{1 \text{ ft}}{12 \text{ in}}\right) = 37.7 \text{ ft}^2$$

$$A_{s4} = \pi D(L_{\text{pile}} - (L_1 + L_2 + L_3)) = \pi(18 \text{ in})(11 \text{ ft})\left(\frac{1 \text{ ft}}{12 \text{ in}}\right) = 51.8 \text{ ft}^2$$

The friction capacity of the entire pile length is the sum of the individual layer friction values throughout the length of the pile.

$$\begin{aligned} Q_s = \sum Q_s &= f_{s1}A_{s1} + f_{s2}A_{s2} + f_{s3}A_{s3} + f_{s4}A_{s4} \\ &= \left(111.4 \ \frac{\text{lbf}}{\text{ft}^2}\right)(23.6 \text{ ft}^2) \\ &\quad + \left(240.3 \ \frac{\text{lbf}}{\text{ft}^2}\right)(28.3 \text{ ft}^2) \\ &\quad + \left(512.1 \ \frac{\text{lbf}}{\text{ft}^2}\right)(37.7 \text{ ft}^2) \\ &\quad + \left(1033.5 \ \frac{\text{lbf}}{\text{ft}^2}\right)(51.8 \text{ ft}^2) \\ &= 82{,}271 \text{ lbf} \quad (82 \text{ kips}) \end{aligned}$$

The answer is (B).

Why Other Options Are Wrong

(A) This option fails to convert the saturated densities given in layers 2, 3, and 4 to effective (buoyant) densities prior to the calculation of effective stress.

(C) This option is incorrect because it uses the angle of internal friction rather than the friction angle in calculating the friction capacity.

(D) This option is incorrect because it takes the cohesion as equivalent to the unconfined compressive strength rather than using half the value in soil layers 3 and 4.

SOLUTION 11

The total (ultimate) pile capacity is the sum of the friction (side) capacity and the end bearing (tip) capacity.

$$Q_{\text{ult}} = Q_f + Q_p$$

The β-method is requested for determination of the friction capacity.

$$Q_f = f_s \pi BL$$

$$f_s = \beta \sigma'_{\text{ave}}$$

For the soft clay layer, the skin friction factor can be determined using the average effective stress and published information for typical values of β.

$$\begin{aligned} f_s &= \beta \sigma'_{\text{ave}} \\ \sigma'_{\text{I,ave}} &= \sigma'_{\text{I,midpoint}} = \sigma_{\text{I,midpoint}} - u_{\text{I,midpoint}} \\ &= \gamma_{t,\text{I}}\left(\frac{L_\text{I}}{2}\right) + \gamma_w D_{\text{harbor}} - \gamma_w D_w \\ &= \left(96.2 \ \frac{\text{lbf}}{\text{ft}^3}\right)\left(\frac{82 \text{ ft}}{2}\right) + \left(62.4 \ \frac{\text{lbf}}{\text{ft}^3}\right)(15 \text{ ft}) \\ &\quad - \left(62.4 \ \frac{\text{lbf}}{\text{ft}^3}\right)(56 \text{ ft}) \\ &= 1385.8 \text{ lbf/ft}^2 \\ \beta &= 0.26 \\ f_{s,\text{I}} &= \beta_\text{I} \sigma'_{\text{I,ave}} = (0.26)\left(1385.8 \ \frac{\text{lbf}}{\text{ft}^2}\right) \\ &= 360.3 \text{ lbf/ft}^2 \end{aligned}$$

For the medium-stiff clay layer, the skin friction factor can be determined using the average effective stress along the length of pile that penetrates the layer.

$$\begin{aligned} \sigma'_{\text{II,ave}} &= \sigma'_{\text{II,mid-penetration}} \\ &= \sigma_{\text{II,mid-penetration}} - u_{\text{II,mid-penetration}} \\ &= \gamma_{t,\text{I}} L_\text{I} + \gamma_w D_{\text{harbor}} + \gamma_{\text{sat,II}} \\ &\quad \times \left(\frac{L_P - (L_\text{I} + 15 \text{ ft})}{2}\right) - \gamma_w D_w \\ &= \left(96.2 \ \frac{\text{lbf}}{\text{ft}^3}\right)(82 \text{ ft}) + \left(62.4 \ \frac{\text{lbf}}{\text{ft}^3}\right)(15 \text{ ft}) \\ &\quad + \left(102.6 \ \frac{\text{lbf}}{\text{ft}^3}\right)\left(\frac{140 \text{ ft} - 82 \text{ ft} - 15 \text{ ft}}{2}\right) \\ &\quad - \left(62.4 \ \frac{\text{lbf}}{\text{ft}^3}\right)(118.5 \text{ ft}) \\ &= 3636 \text{ lbf/ft}^2 \\ \beta &= 0.3 \\ f_{s,\text{II}} &= \beta_{\text{II}} \sigma'_{\text{II,ave}} = (0.3)\left(3636 \ \frac{\text{lbf}}{\text{ft}^2}\right) \\ &= 1090.8 \text{ lbf/ft}^2 \end{aligned}$$

The friction capacity can be calculated.

$$\begin{aligned}Q_f &= f_{s,\text{I}}\pi BL_{\text{I}} + f_{s,\text{II}}\pi BL_{\text{II}}\\ &= \left(360.3\ \frac{\text{lbf}}{\text{ft}^2}\right)\pi(12\text{ in})(82\text{ ft})\left(\frac{1\text{ ft}}{12\text{ in}}\right)\\ &\quad+\left(1090.8\ \frac{\text{lbf}}{\text{ft}^2}\right)\pi(12\text{ in})(43\text{ ft})\left(\frac{1\text{ ft}}{12\text{ in}}\right)\\ &= 240{,}172\text{ lbf}\quad(240.2\text{ kips})\end{aligned}$$

The end bearing location is 43 ft below the top of layer II. The total pile tip capacity can be calculated. N_c is typically taken as 9 for driven piles.

$$c = \frac{q_u}{2}$$

$$\begin{aligned}Q_p &= N_c c A_p = 9\left(\frac{q_u}{2}\right)A_p\\ &= (9)\left(\frac{2010\ \frac{\text{lbf}}{\text{ft}^2}}{2}\right)\left(\frac{\pi}{4}\right)(12\text{ in})^2\left(\frac{1\text{ ft}}{12\text{ in}}\right)^2\\ &= 7100\text{ lbf}\quad(7.1\text{ kips})\end{aligned}$$

The total (ultimate) and allowable pile capacities can both be calculated.

$$\begin{aligned}Q_{\text{ult}} &= Q_f + Q_p = 240.2\text{ kips} + 7.1\text{ kips}\\ &= 247.3\text{ kips}\\ Q_a &= \frac{Q_{\text{ult}}}{F} = \frac{247.3\text{ kips}}{3}\\ &= 82.4\text{ kips}\quad(82\text{ kips})\end{aligned}$$

The answer is (C).

Why Other Options Are Wrong

(A) This incorrect solution omits the depth of water in the effective stress calculation.

(B) This option is incorrect because it calculates the pile tip capacity using the cohesion value and omits the depth of the water in the effective stress calculation. The end bearing value is determined using the unconfined compressive strength value without typical conversion by dividing by 2.

(D) This solution fails to reduce the ultimate value by the factor of safety and omits the depth of water in the effective stress calculation.

SOLUTION 12

The structural loading is uniform and concentric. Therefore, neither moment nor lateral load is transferred to the pier group. The ultimate block action capacity is the sum of the friction (side) capacity and the end bearing (tip) capacity.

$$Q_{\text{ult}} = Q_s + Q_p$$

The ultimate individual pier capacity is one fourth that of the overall ultimate group capacity.

$$Q_{\text{ult,pile}} = \frac{Q_s + Q_p}{4}$$

Keeping in mind that cohesion, c, is typically taken as $q_u/2$, the friction capacity for block action can be calculated.

$$\begin{aligned}Q_s &= 2(b+w)Lc\\ &= 2(b+w)\left(L_{\text{A}}c_{\text{A}} + L_{\text{B}}c_{\text{B}} + L_{\text{C}}\left(\frac{q_{u,\text{C}}}{2}\right) + L_{\text{D}}\left(\frac{q_{u,\text{D}}}{2}\right)\right)\\ &= (2)(5.5\text{ ft} + 5.5\text{ ft})\left(\begin{array}{l}(3\text{ ft})\left(200\ \frac{\text{lbf}}{\text{ft}^2}\right)\\ +(6\text{ ft})\left(200\ \frac{\text{lbf}}{\text{ft}^2}\right)\\ +(10\text{ ft})\left(\frac{600\ \frac{\text{lbf}}{\text{ft}^2}}{2}\right)\\ +(6\text{ ft})\left(\frac{700\ \frac{\text{lbf}}{\text{ft}^2}}{2}\right)\end{array}\right)\\ &= 151{,}800\text{ lbf}\quad(152\text{ kips})\end{aligned}$$

The end bearing capacity for block action can be calculated. N_c is typically taken as 9.

$$\begin{aligned}Q_p &= N_c c_D bw = 9\left(\frac{q_{u,D}}{2}\right)bw\\ &= (9)\left(\frac{700\ \frac{\text{lbf}}{\text{ft}^2}}{2}\right)(5.5\text{ ft})(5.5\text{ ft})\\ &= 95{,}288\text{ lbf}\quad(95\text{ kips})\end{aligned}$$

The ultimate block action capacity, both for the group and individually, can be determined.

$$Q_{\text{ult}} = Q_s + Q_p = 152 \text{ kips} + 95 \text{ kips} = 247 \text{ kips}$$

$$Q_{\text{ult,pile}} = \frac{247 \text{ kips}}{4} = 61.8 \text{ kips} \quad (62 \text{ kips})$$

The answer is (C).

Why Other Options Are Wrong

(A) This option mistakenly calculates the allowable individual pier capacity.

(B) This option mistakenly calculates the individual pier capacity using the β-method when the problem statement specifies the assumption of block action behavior.

(D) This option fails to calculate the individual pier capacity from the pile.

SOLUTION 13

The total (ultimate) pile capacity is the sum of the friction (side) capacity and the end bearing (tip) capacity.

$$Q_{\text{ult}} = Q_f + Q_p$$

The α-method is requested for determination of the friction capacity.

$$Q_f = f_s \pi BL$$
$$f_s = \alpha c$$

The skin friction factor for each soil layer can be determined and summarized. The value for cohesion is typically taken as half the unconfined compressive strength value.

layer	unconfined compressive strength, q_u (lbf/in²)	cohesion, c (lbf/in²)	cohesion, c (lbf/ft²)	α (ave)	skin friction factor, f_s (lbf/in²)
1	–	0.5	72	1	72
2	–	0.5	72	1	72
3	7	3.5	504	1	504
4	14	7	1008	0.83	837

The surface area of the pile passing through each soil layer can be calculated and summarized in a similar manner.

layer	pile side width, B (ft)	length of pile, L (ft)	surface area, $4BL$ (ft²)
1	1.2	8	38.4
2	1.2	10	48.0
3	1.2	15	72.0
4	1.2	12	57.6

The total friction capacity can be calculated and tabulated.

layer	skin friction factor, f_s (lbf/ft²)	surface area, $4BL$ (ft²)	friction capacity, f_s4BL (lbf)
1	72	38.4	2765
2	72	48.0	3456
3	504	72.0	36,288
4	837	57.6	48,211
total friction capacity, Q_f (lbf)			90,720

The end bearing location is in layer 4. The total pile tip capacity can be calculated. N_c is typically taken as 9 for driven piles.

$$\begin{aligned} Q_p &= N_c c A_p \\ &= (9)\left(1008 \frac{\text{lbf}}{\text{ft}^2}\right)(14 \text{ in})(14 \text{ in})\left(\frac{1 \text{ ft}}{12 \text{ in}}\right)^2 \\ &= 12{,}348 \text{ lbf} \end{aligned}$$

The total (ultimate) pile capacity can be calculated.

$$\begin{aligned} Q_{\text{ult}} &= Q_f + Q_p = 90{,}720 \text{ lbf} + 12{,}348 \text{ lbf} \\ &= 103{,}068 \text{ lbf} \quad (100 \text{ kips}) \end{aligned}$$

The answer is (C).

Why Other Options Are Wrong

(A) This option fails to convert the pile dimensions from inches to feet, causing the units to be wrong throughout the entire solution.

(B) This option mistakenly assumes the given side width value is the pile diameter and subsequently uses the equations associated with a circular pile. The skin friction factor for each soil layer is tabulated.

(D) This incorrect option takes the cohesion as equivalent to the unconfined compressive strength rather than using half the value. The skin friction factor for each soil layer can be tabulated.